W9-ABI-262

Understanding the
Alcoholic's Mind

Understanding the Alcoholic's Mind

The Nature of Craving and How to Control It

ARNOLD M. LUDWIG, M.D.

New York   Oxford
OXFORD UNIVERSITY PRESS
1988

Oxford University Press

Oxford   New York   Toronto
Delhi   Bombay   Calcutta   Madras   Karachi
Petaling Jaya   Singapore   Hong Kong   Tokyo
Nairobi   Dar es Salaam   Cape Town
Melbourne   Auckland

and associated companies in
Beirut   Berlin   Ibadan   Nicosia

Published by Oxford University Press, Inc.,
200 Madison Avenue, New York, New York   10016

Oxford is a registered trademark of Oxford University Press

Library of Congress Cataloging-in-Publication Data
Ludwig, Arnold M.
  Understanding the alcoholic's mind.
  Bibliography: p.
  Includes index.
  1. Alcoholism.   2. Alcoholism—Psychological aspects.
I. Title.   [DNLM: 1. Alcoholism—psychology. WM   274 L948u]
RC565.L83   1988   616.86'1   87-11328
ISBN 0-19-504878-4

10 9 8 7 6 5 4 3

Printed in the United States of America on acid-free paper

For Lucinda and Dan

Preface

Many books have been written about why people become alcoholics. Few, of a clinical nature, have been written on how alcoholics manage to recover.

When I became interested in the treatment of alcoholism over twenty-five years ago, I thought I had all the answers. I believed, for example, that if alcoholics truly understood why they drank and what harm drinking was doing to them, they would no longer want to drink. All that was necessary was for them to become educated about their problem. My naiveté was profound. I soon came to realize that I knew little about the real reasons individuals felt compelled to drink. I knew even less about how some, who had once been considered hopeless drunkards, managed to recover.

Over the course of the years, I have tried to remedy my ignorance by interviewing, studying, treating, or interacting with over 1,000 alcoholics from all walks of life and within different settings—outpatient facilities; state, Veterans Administration, university, private, and community hospitals; Alcoholics Anonymous meetings; detoxification centers; halfway houses; informal gatherings; public meetings or private homes. About one fourth of these individuals had quit drinking alcohol for significant periods of time. I have also had the opportunity to keep abreast of most of the relevant research and literature on treatment and recovery. From these varied experiences, I have come to several not-so-surprising conclusions:

- Despite the discouraging results of most alcoholism treatment programs or efforts at self-cure, many alcoholics, under the proper circumstances, do manage to recover.
- Those who recover do so by adopting certain characteristic ways of thinking and behaving.
- The changes associated with recovery, if they are to be lasting, tend to involve not only drinking but all aspects of their lives.

The purpose of this book is to elaborate on these conclusions by describing what takes place during the process of recovery. Its basic message is optimistic. The fact that many alcoholics manage to remain sober and lead productive and gratifying lives, even against the odds, offers hope to all. Alcoholism need not always be an irreversible, progressively deteriorating affliction. The potential for recovery is always there. It is simply a matter of knowing what to do.

The audience for this book is anyone with an interest in alcoholism in general and recovery in particular. Certain readers may approach the material from a clinical perspective to increase their knowledge about alcoholism or improve their therapeutic skills; others may find practical ways to apply it for their own personal benefit or "cure." The body of the text is basically descriptive and anecdotal, appropriate for laypersons and professionals alike. Material of a more technical or theoretical nature is included in extensive footnotes at the end of the book for those who wish further details.

One word about style. Throughout the book I use the term "alcoholic" to refer to individuals suffering from alcoholism. Because of its dehumanizing implications, this term is unfortunate, but for stylistic and conventional purposes, it will have to do. I know of no other designation that conveys so much meaning.[1,2]

Lexington, Kentucky   A. M. L.
April 1987

Acknowledgments

In writing this book, I have benefited immensely from the input of many colleagues and friends with a special expertise in the field of alcoholism. In particular, I am indebted to Donald W. Goodwin, M.D., Gordon L. Hyde, M.D., Harold Kalant, M.D., Ph.D., Roger E. Meyer, M.D., Jacquiline Nelson, Margarita Reeves, R.N., David R. Rudy, Ph.D, Verner Stillner, M.D., and Robert Straus, Ph.D., for the time and effort they expended in reviewing an earlier draft of my book and their invaluable suggestions. I also am grateful to my editor, Shelley Reinhardt, for her ongoing support of this project and arranging to have such outstanding consultants as Margaret Bean-Bayog, M.D., Gloria K. Litman, Ph.D., and Alan G. Marlatt, Ph.D. review my manuscript.

Contents

Understanding the Alcoholic's Mind

1

Paradoxes and Contradictions

There is no general agreement about the nature, cause, or treatment of alcoholism. What is an alcoholic? Where does one draw the line between problem drinking and alcoholism, between alcohol dependence and addiction? Is alcoholism one disorder or a collection of different disorders? Is it a moral failing, a bad habit, or a disease? Do alcoholics have distinctive personality features? Is alcoholism hereditary or learned? Does excessive drinking represent a symptomatic expression of an underlying conflict or is it the primary problem itself? Which treatment approach, if any, is most effective? Who is best qualified to help? The questions can go on and on. There are no scientific answers. In the absence of facts, opinions and beliefs tend to prevail. Sometimes they seem paradoxical and inconsistent. Consider the following.

• "Hitting bottom" is presumed to be a necessary step for recovery, even though being in dire straits, for all other illnesses, usually indicates a poor rather than favorable prognosis.
• Recovery from alcoholism—at least according to Alcoholics Anonymous—depends upon an admission and acceptance of helpless-

ness by the alcoholic. To gain control over the disease, the individual must relinquish his personal will to that of a Higher Power and to the community of fellow alcoholics. In effect, the acceptance of personal weakness becomes the basis of strength.

• Along with other addictions, alcoholism is unique in the extent to which the individual is blamed if the treatment fails. If the alcoholic does not remain abstinent, therapists and staff presume that he is unmotivated for or unreceptive to help.

• In many hospital treatment settings, alcoholics are immediately discharged from the program if they are presumed to be uncooperative, unmotivated, setting poor examples for others, or if they are found to be intoxicated or drinking on the premises. In other words, they are not regarded as suitable for treatment if they show evidence of their sickness: namely, an inability to control their drinking. The catch-22 is that they must remain sober in order to receive help.

• Alcoholism represents a disorder in which alcoholics presumably cannot refrain from drinking, yet most therapeutic programs expect them, once abstinent, never to consume that first drink.

• Alcoholics are regarded as "sick"—at least for purposes of hospitalization or treatment—but society tends to hold them responsible for their transgressions or crimes.

• Alcoholism as a disease is still presumed to exist even when an alcoholic has been sober for years. Once an alcoholic, always an alcoholic, as the saying goes. Even with cancer, the prognosis is not that grim.

• Because alcoholism is regarded as a "disease", certain therapeutic agencies do not hold alcoholics responsible for the harm caused by past drinking, but they do regard them as responsible for their present and future behaviors, an important and interesting distinction.

• Alcoholism is the only "disease" for which communion with a Higher Power is regarded by many as an essential element for recovery. Even physicians who are otherwise wary of the role of religion in medicine and rely primarily on drugs and procedures of proven, scientific merit for the treatment of serious illnesses, endorse participation in Alcoholics Anonymous, which has a strong spiritual emphasis, as an important component of therapy.

• Alcoholism is a "disease" in which characteristic symptoms, such as urges and cravings to drink, can appear mysteriously at certain times, for example, during evenings and weekends, and be absent at others, as at work or at church. With the exception of other addictions, what medical diseases are so dependent upon the mental expectations of the sufferers and the physical settings in which they exist?

• When uncomplicated by medical problems, alcoholism is assumed by many to be better treated by those who have suffered from it than by trained professionals who never have. Alcoholism is the only "disease" about which the recovered patient is presumed to know more than the doctor.

• Alcoholism is the only disorder for which the label of "alcoholic" is regarded as a stigma or a moral condemnation by some and a badge of honor by others, the qualification for automatic membership into a fellowship of like-minded sufferers. Unlike individuals with cancer, hypertension, or other medical diseases, alcoholics tend to become defined by their disease, which presumably obliterates their individuality, putting them on the same footing as all other alcoholics. They are alcoholics first and foremost, even when sober, rather than individuals with a drinking problem.

With all these paradoxical claims, it is not surprising that so many individuals claim expertise in this area. Unlike most medical disorders, for which only certain specialists are acknowledged as experts, alcoholism has no such restraints. Sociologists, social psychologists, clinical psychologists, social workers, pharmacologists, public health officials, psychiatrists, internists, ministers, priests, law enforcement authorities, politicians, and recovered alcoholics themselves tend to view alcoholism from their own special perspectives—*moral, biological, psychological, social, spiritual,* or some combination thereof—each of which contains a different set of assumptions about the nature of alcoholism and its prevention, management, or treatment.[1]

With the *moral perspective*, alcoholism is regarded as an expression of immorality, a moral failing, or a vice. Self-indulgence, hedonism, gutlessness, irresponsibility, laziness, and moral turpitude are supposedly at the root of the problem. Drinking represents a sign of weakness, requiring the exercise of willpower for cure. If alcoholics feel guilty about their drinking, they should, since it's their fault they are in that position to begin with.

This is the attitude that much of society adopts toward alcoholics. Drunk tanks, jail, convictions for drunk and disorderly conduct, and the general opprobrium with which alcoholics are regarded in many quarters—even by physicians and health-care professionals—tends to imply they are "bad."

Alcoholics themselves are not immune from these same attitudes. Self-recriminations, guilt, despair, and self-disgust tend to follow in the wake of a drinking bout, often leading to renewed promises to quit. In fact, it is this moral posture, these assumptions about personal responsibility and willpower, that often underlies the impetus for recovery, the reason that many alcoholics decide to go "on the wagon" or "take the pledge." Unable to live with themselves anymore, they opt to quit drinking.

With the *biological perspective*, alcoholism is regarded as a "disease," a result of disordered body chemistry, allergy, nutritional deficiency, physical vulnerability, or genetic inheritance. Alcoholics do what they do out of physiological necessity rather than choice. They are "sick," and as such, cannot be held responsible for their past actions.

Because alcoholism is presumed to be due to physical causes, physical procedures are of course prescribed for its "cure." Typical treatments involve the use of high doses of certain vitamins to correct presumed nutritional deficiencies, aversive conditioning procedures with drugs like Antabuse or emetine, which can produce nausea and gagging in response to the presence or even the thought of liquor, or antidepressants or other mood-modifying agents to correct any underlying psychiatric disturbance.

With the *psychological perspective*, alcoholism is regarded as largely a learned disorder or a maladaptive behavior pattern, representing either a way of coping with problems, a symptomatic expression of a deep-seated emotional conflict, a reflection of irrational attitudes and assumptions, a series of conditioned responses, or simply a "bad habit." The remarkable ability of alcohol to provide temporary relief from emotional distress or to remove social inhibitions, among other attributes, presumably reinforces the compulsive need to drink.

This perspective regards alcoholism as neither a moral failing nor a disease. It is not a vice since it is not something that the individual deliberately or consciously brings on himself. The alcoholic drinks as he does because of what happened to him in the past, because of the way he has learned to think and behave, because he has been powerfully conditioned to do so, because of a poor self-image, because of an underlying tension or depression, or even because of a mistaken notion that he needs to drink and lacks control over his urge. But alcoholism is also not regarded as a disease since it is not believed to be due to physical causes; only attitudes, feelings, and behaviors need be taken into account. Therapeutic approaches based on these assumptions about etiology tend to emphasize the need for alcoholics to gain insight into their motivations, feelings, and behavior, to learn to recognize and avoid cues that stimulate craving, to alter faulty or irrational attitudes that foster drinking, to be more assertive, to acquire more effective coping skills, or to act in more self-enlightened ways.

With the *social perspective*, alcoholism is presumed to be due to environmental forces impinging upon the individual. No alcoholic is completely autonomous or independent of others or the context in which he lives. It is difficult, for example, to be an alcoholic if alcohol is never available or if one lives in a culture in which its use is punished or its cost is prohibitive. Because of this, the behavior of an alcoholic can only be understood in the context of a broader social network—his or her marriage, family, community, and society.

Liquor taxes, local laws against the sale of alcoholic beverages, incarceration or fines for public intoxication, and a host of other measures are ways that society has tried to regulate alcohol consumption. Within a more limited, therapeutic framework, the focus has been on family dynamics. Alcoholism, for example, is not considered to be the problem of the designated alcoholic as much as of the entire, immediate family. The alcoholic merely serves as a scapegoat, a repository or funnel for the emotional pathology of others. That is his or her assigned role to play, much as it was Mary Tyrone's role to play the drug addict and bearer of the family secret in Eugene O'Neill's *Long Day's Journey Into Night*. It is the role of other family members to act out their complementary parts as martyred spouse, neglected child, or disgraced parent. By unconsciously engaging in these interactional "games," the alcoholic helps to preserve a family equilibrium necessary for its maintenance as an intact unit.

Naturally, with this outlook, the treatment of alcoholism is more likely to be effective if the pathology of the entire family is brought to light so that new, healthier interactional patterns can take place. Couples therapy, family counseling, spouse groups, and, to a certain extent, Al-Anon meetings are partly predicated upon his premise. Correct the interpersonal problems and the alcoholism is more likely to improve.

And last, with the *spiritual perspective*, alcoholism is regarded as a sign of a person's alienation from a Higher Power or divine being. It is the sin of pride that allows the alcoholic to believe that he is capable of exerting control over something he has no control over—his drinking—and that makes him so vulnerable to temptation. It is only when the alcoholic no longer can maintain the fiction of being in control, when stark reality begins to shatter his illusions, that he is able to set aside his false pride and turn outside of himself for support and guidance. This is potentially a time of revelation, of a radical reorganization in values, and of communion with a force greater than himself—a power that can take the form of a personal God, a natural law, a religious community, or a fellowship of like-minded sufferers. Only by overcoming this rift between himself and this Higher Power can the alcoholic ever expect to be healed. This is one of the major premises of organizations such as Alcoholics Anonymous (AA).

What these different perspectives reveal about alcoholism is that it is a complex condition, with each perspective providing only part of the picture. The paradoxes and contradictions already noted come into existence when two or more of these seemingly incompatible or mutually exclusive views are combined. Logically, combinations of perspectives, like mixed metaphors, should not be proper, but, realistically, they tend to capture the true nature of alcoholism better than any single perspective. Alcoholism may be a disease, for example, but that does not relieve alcoholics of the responsibility for their actions. Once they start drinking, alcoholics may have trouble stopping, but they do have the power to decide whether or not to take the first drink. Self-indulgence may have led many alcoholics to imbibe, but they may also be more genetically predisposed than nonalcoholics to drink excessively. Alcoholics may be held personally responsible for their behavior, but that does not mean they are immune from family dynamics that keep casting them in a deviant social role. Alcoholics may desperately want help, but that may not keep them from fooling themselves or conning others when they are in treatment. Alcoholics may not stop drinking, but that does not mean they are unmotivated to get better. Many alcoholics may recover entirely on their own initiative, but that does not mean that others cannot benefit from spiritual sources of help. These examples of overlapping or complementary perspectives go on and on.

What are the implications of all this for alcoholics seeking help? With so many perspectives on alcoholism available, any explanation of what takes place during the process of recovery is likely to be both confusing and deficient unless all of them can be linked together and a common area of overlap found. This common area can then allow a common language to be used, a lingua franca of sorts—one stripped of any doctrinaire trappings—to describe, in a coherent and more comprehensive way, what individuals need to do to insure sobriety.

But where to look for this area of overlap?

The answer seems to be . . . *in the mind.* It is there that alcoholics conjure up reasons and excuses to drink. It is there that they first entertain decisions to quit, struggle with their daily urges, and make choices about the future direction of their lives. It is there that they deceive themselves and others. It is there that they acknowledge their helplessness, seek spiritual support, or turn to others for help. If they are to remain sober, they somehow must get to the point, regardless of whatever perspective of their problem they or others hold, where they not only can envision living without alcohol but actually prefer doing so. For that to happen, their thoughts, attitudes, and motivations have to change. It is my plan to describe how this can come about.

2

The Lure
of the Sirens

Exhausted from the long period of cramming for his exams, Jack London yearned to set sail on the open seas, for a chance to put his mind at rest and his body to work. "Benicia showed before me . . . where in the old days I had lived and drunk deep. I had no intention of stopping at Benicia," he wrote. "The tide favored, the wind was fair and howling—glorious sailing for a sailor. . . . And yet, when I laid eyes on those fishing arks lying in the waterfront tules, without debate, on the instant, I put down my tiller, came in on the sheet, and headed for shore. On the instant, out of the profound of my brain-fog, I knew what I wanted. I wanted to drink. I wanted to get drunk."[1]

The accounts of so many other alcoholics, like Jack London's, also reflect the inevitable "bad luck" of just happening to find themselves in compromising situations, sometimes after months or years of abstinence, in which the urge to drink becomes irresistible. Despite their best intentions to remain sober, fate perversely conspires to place obstacles in their paths—or so they perceive their star-crossed lives.

But are these occasions entirely accidental, or have the individuals somehow managed to bring them about? Like Odysseus, they lash themselves to the mast in the hope of avoiding the lure of the Sirens

. . . but they forget that it was they who set sail in the perilous waters in the first place. If they truly had been motivated to remain sober, then why did they not simply remain ashore?

What this illustrates is that the alcoholic's worst enemy is not the bottle or bad luck but his own mind, within which is the ever-present Trojan horse of desire, waiting to smuggle in the enemy when the defenses have been lulled into complacency. What must be recognized is that in this case the brain is much less an organ of rationality than of rationalization. Because the individual has to live with himself, his mind tries to legitimatize his intentions and behaviors so that he need not feel guilty, so that he can convince himself that he is really making the right choice. This is the false logic of John Barleycorn, the deceptive, seductive reasoning, the self-serving excuses and selective denial of reality that Jack London so colorfully writes about. It is not that the individual really wants to drink, so he protests, "he deserves it." He's been working hard lately and "owes himself a drink." Just a touch of whiskey, "to keep off the chill" and "to settle the stomach." Or "just one for the road." And as long as he's going to have a drink, he might as well "put a couple under his belt" and come home drunk since his wife will complain anyhow. Or, if he is like Elmore Leonard, the novelist, he can try disguising the taste of his whiskey with an unfamiliar soft drink to help him deny the fact that he's actually drinking hard liquor.[2]

But these rationalizations are not only used to justify drinking. Often, the mind of the abstinent alcoholic is so devious as to come up with a number of seemingly innocuous decisions, sometimes beginning days, weeks, months, and even years in advance of a lapse, which eventually place the individual in a situation in which a return to drink becomes inevitable.[3] Through a complicated, often intricate series of mental maneuvers, designed mainly to outwit himself—to slip his real but perhaps unconscious intentions past the gullible sentry in his mind— the individual unwittingly ensures that he will be seduced into drinking. After getting himself into a position to be tempted, the alcoholic, like Oscar Wilde who could resist anything but temptation, inevitably succumbs. But, if he could be really honest with himself, he would recognize that his lapse was not because of bad luck or fate but because, at some level of awareness, he planned to do this all along. He was unconsciously responsible for his own downfall. The covert agenda had been formulated long ago. The problem was how to implement it without its seeming that the decision for relapse was both conscious and planned.

In a sense, the alcoholic keeps tossing out banana skins which he "accidentally" manages to step on and then slip. His brain keeps coming up with ingenious booby traps and inadvertent pitfalls, arranging for the bad luck to occur. Among the scores of alcoholics I have known, I continue to be amazed at how reasonably intelligent individuals can actually believe their claims of innocence and not recognize the blatant rationalizations, selective perceptions, distortions of reality, denial, and self-deception in their slips.

One man "accidentally" bought apricot wine from a drugstore, swearing that he thought it was juice, and then sampled it at home.

One woman, who developed a sudden passion for baking cakes, began nipping on the vanilla extract because she believed it essential to taste all the ingredients before they were added.

Tired after a long night of driving, one individual pulled into the motel and, as he had always done, automatically filled the empty bucket with ice. He poured himself a glass of water, took a swallow, and then gagged at the chlorinated taste. There was only one remedy that he could think of. He got the flask of scotch from the glove compartment of his car (now remembering that he had forgotten to remove it when he took the pledge months ago) and added it to the water. But only because he was "thirsty," or so he told himself.

After several years of abstinence, one man got religion and began taking Communion with wine, regarding it as safe since it was the body of Christ. Within several weeks, he was drinking heavily again.

On his way home from work, one individual figured that the snarled traffic ahead indicated an accident, so he took a long detour, which carried him several miles out of his way. By accident, he happened to turn on the street of a once-favorite bar that he had not frequented in years. Noting that his radiator seemed to be overheating, he thought it best to turn off the engine and just pass some time inside while it cooled off. Instead of the intended soft drink, he ordered a beer.

One individual developed a persistent cough after he took up smoking again. Rather than cut back on his two- to three-pack-a-day habit, he began to rely on an elixir of terpin hydrate as a palliative for the cough—an elixir that is 40 percent alcohol. The cough, naturally, continued, forcing him to use greater and greater quantities of the elixir to provide sufficient relief. He did so despite the fact that there are numerous other cough medicines with no alcohol.

One man swore he ordered ginger ale on the plane, but the stewardess accidentally poured him champagne.

After months of taking Antabuse, a drug that can make people extremely sick if they also drink alcohol, one individual found himself feeling more and more ill. Rather than consult his physician, he concluded that the medication did not agree with him, so he stopped taking it. Several days later, for some unknown reason, he found that his hands were shaking and his body felt trembly inside. There was only one proven remedy for that, at least by past experience—a drink.

Then there are the scores of alcoholics, too many to estimate, who quit taking Antabuse because they have their problem "licked" but who, because of "bad luck," their old nemesis, manage to slip within a week or two.

There are also the many recovering alcoholics who manage to find jobs only as bartenders, cocktail waitresses, or liquor store clerks, those who need to stock up on liquor in case friends stop over and then proceed to invite over their heaviest drinking friends, those who prefer to drink their coffee in bars, or those who become social butterflies and flit around from party to party sipping on ginger ales or virgin marys. Hadn't any of these individuals been told that one of the best ways to resist temptation is to avoid it?

In all my years of going to parties, I never remember picking up someone else's drink or having someone substitute an alcoholic beverage for my soft drink when I had one, yet countless, formerly sober alcoholics inform me about their bad luck in accidentally picking up someone else's alcoholic drink and consuming it before they could catch themselves or of having had someone switch a drink on them.

One of the most creative buildups for a drunk was described by an individual who, for weeks before his relapse, began finding more and more faults with his wife. The arguments became so heated and the tension so great, that he believed he had only two choices: to physically beat her or to find some relief with alcohol. Opposed to violence, he opted for the alcohol.

Other insightful individuals have informed me that they could now see in retrospect how they would seek out or generate stress in their lives, sometimes even pushing themselves to the breaking point, so that they would have an excuse to drink. It was not that they wanted to get drunk, so they believed at the time, but only that they had no other alternative.

One individual wryly summed up the general plight of the recovering alcoholic when he remarked, "If you are an alcoholic, the lawn mower won't start unless you add ethyl alcohol—not to the gasoline but to your belly."

There is an AA saying, "If you don't want to slip, avoid slippery places." There is wisdom in this saying, but the problem for the alcoholic is to identify these slippery places or situations, especially when a clever, devious homunculus is at work in his mind trying to outfox him. Generally, the recovering alcoholic has no trouble recognizing the more obvious, dangerous situations, like noisy bars, raucous cocktail parties, two-hour martini lunches, or a six-pack of cold beer with the boys, and can mobilize his resolve to resist these patent opportunities to drink. The choice is there, right out in the open, with no way to deceive himself if he takes a drink. The real danger occurs, though, when the situation is more subtle, when the risk seems remote, or when the circumstances appear innocent. Surely, no harm can come from sipping on a glass of vintage wine during supper, while seated across the candlelit table from an attractive, charming date. What can be more romantic? Or what can be more innocent than belting down a couple of scotches to settle your nerves while traveling by plane? How else can you deal with your flying phobia? And what about that bottle of expensive champagne for the special occasion, such as a twenty-fifth wedding anniversary, the celebration of a birth, the landing of a lucrative contract? The point is that the recovering alcoholic, in order to avoid or minimize guilt or the shame of being weak-willed, must unwittingly find some face-saving way to lapse so that it will seem that he really had no other choice. Or that he was tricked into doing it. Or that what he did seemed legal, appropriate, or normal at the time. What this indicates is that the alcoholic who professes a desire not to drink is free to arrange his downfall only if he can disguise all his machinations from himself.

The question arises about why a recovering alcoholic, who, at least consciously, wants to remain sober and has been reasonably successful in doing so for some time, should be covertly maneuvering to arrange a slip, to undo all that he has so far accomplished. Sure, there is the powerful appeal of the intoxication itself, the attraction of experiencing "the place where it's all O.K., where body and soul are one."[4] But there has to be something more, something to justify doing what he knows he shouldn't do, particularly after having established that he can get along perfectly well without alcohol.

One of the main reasons that many relapse-prone alcoholics continue to be plagued by the prospects of drinking, why they end up beguiling themselves into doing what they unwittingly want to do, is that lurking somewhere within their minds, despite conscious protestations to the contrary, is the belief that they have at least one drunk left, that they can survive another bout of extended intoxication, if not with impunity then without irreparable consequences, all past evidence to the contrary.[5] This belief is often unconscious, and if it does become conscious, they dare not verbalize it since it is antithetical to successful recovery. "I remember him well," Mark Only wrote,[6] "the person I was as an active alcoholic. His ghost or shadow or incubus lurks there within me"—and, I might add, whispering in his ear and twisting his thoughts, always panhandling for another drink, just one more fling. It is not the thought of quitting drink that bothers alcoholics, since they have imagined that thousands of times, but the notion of never drinking again. That is almost incomprehensible. Surely, at some future time they can risk it—so they tell themselves—when their lives have been set in order, when their thoughts have been straightened out, when they are wise enough to avoid the mistakes of their past drinking bouts, when they no longer crave alcohol, when medical science comes up with a new cure, and so on. It is also this hope that has them cling to the prospects of "social drinking,"[7] to controlling their drinking by new approaches, like shifting from liquor to wine, diluting the alcohol with a mixer, drinking only at specified times, taking sips rather than gulps, or a thousand other changes that will permit them to drink because "the next time it will be different," since they want to believe they have learned from their mistakes. It is this unstated, usually unrecognized expectation about their drinking that often keeps the embers of desire alive and maintains their vulnerability. If alcoholics truly believe that under *no circumstances*, at *no time*, at *no place*, and with *no exceptions* do they have the option of returning to drink, then the voice of seduction will be completely stilled. Individuals cannot be tempted to do something that no longer is tempting.

Let's now examine some less obvious ways that alcoholics set themselves up for a relapse.

3

The Dry Drunk

Carl felt shabbily treated by the president of the bank. For no apparent reason, other than spite, the president stripped him of his position as loan officer and put someone with less seniority in charge. The more he thought about what happened, the angrier he got. He began withdrawing at work. He began having trouble sleeping at night. He became irritable with his wife and children. His financial judgment began to suffer. He impulsively bought an expensive sports car and made several bad financial investments. The more he brooded about his job, the more convinced he became that he was being wronged. Although he had been sober for over five years and had been active in Alcoholics Anonymous, his wife became alarmed that he might be drinking again, since this was how he had acted during those earlier terrible years. But whenever she voiced her concerns to him he exploded, accusing her of not trusting him and making matters worse.

Had Carl been secretly tippling, or had the wife misconstrued his behavior? Actually, he had been sober this whole time. But that did not mean there was no cause for her concern. In some ways, his behavior was very similar to what it was when he had been drinking in the past. He had not been drinking, but it was only a matter of time before he was.

This type of behavior in alcoholics has come to be known as a "dry drunk" because it is often indistinguishable from that shown during a buildup to a drinking spree[1] and recreates the same kind of thought processes and associated feelings that alcoholics either try to suppress or stimulate through drinking. Typical of their thought processes during the buildup are tendencies to feel sorry for themselves, to blame others for whatever goes wrong, to nurse grievances, to become preoccupied with petty concerns, to dwell on the past, to keep imagining the worst, to feel alienated from others, to shirk responsibilities, to overreact to frustrations, to act impulsively, and to become obsessed with immediate pleasures. All the ingredients for a drinking bout are present, but individuals may not recognize that at the time.

But because people think this way does not mean that relapse is inevitable. Though the likelihood of drinking again is high, many individuals can remain in this unstable state of mind for months or years and be sober. Fortunately for Carl, he finally was able to recognize just how dangerous his thinking had become, and so made arrangements to go to a special AA-sponsored program for recovering alcoholics. After one week of reflection, soul-searching, and intensive discussion with others, his anger was gone and he felt spiritually renewed.

What all this indicates is that recovering alcoholics need to beware of a state of mind that can predispose them to a relapse—a state of mind characterized by certain dangerous attitudes and thought processes, aptly described as "stinking thinking." The reason these thought processes are dangerous is because they induce alcoholics to relax their guard. Why worry about relapse when the craving for alcohol may be absent and there are clearly real problems underlying their thoughts? But what they fail to recognize is that the danger is not in their thoughts at the moment but in what they can later spawn. Stinking thinking, as the saying goes, leads to drinking thinking, and then, usually, to drinking without thinking.

That is not to say that this kind of thinking is always pathological or limited only to alcoholics. In a broader context, it qualifies as being "neurotic," "immature," or "irrational" and occurs in most people at one time or another.[2] However, it is not the thinking itself that creates the problem, but how people deal with it. If alcoholics can learn to dismiss this thinking from their minds whenever it appears, recognize it for what it is, or counter it with contrary thoughts, it need not be too disruptive. It is only when this thinking persists, when individuals give it credence and regard it as a reflection of the true state of affairs, and when their usual outlets and constructive ways of dealing with it are blocked, that difficulties can be expected to arise and the attraction of intoxication to grow.

From my own studies, which involved detailed, in-depth interviews with numerous alcoholics in various stages of recovery, I have been able to identify nine overlapping patterns of thoughts and attitudes that seem to predispose to drinking. What these thought processes represent are private self-statements, a type of nonvocal inner speech that usually, but not always, serves as a mediator between intentions and deeds. In a sense, these different thought patterns qualify as "scripts" since, like lines in a play, alcoholics tend to learn them by heart, often rehearsing them over and over in their minds until they can convincingly throw themselves into the part. But what is so remarkable about these scripts is how hackneyed or stereotyped they tend to be. Individuals from all walks of life seem to employ the same kinds of themes—with allowances, naturally, for extemporaneous and personal variations—indicating that the same sorts of excuses are almost universally employed. Some individuals rely on only one or two scripts, keeping themselves vulnerable by mentally repeating these lines, while others, like repertory players, become proficient with many scripts, creatively managing to stay emotionally off-balance. But despite the different themes represented in these thoughts, they all essentially portray the same kind of role: a recovering alcoholic in serious danger of relapse.

Here are descriptions of these various kinds of thought.[3]

With the *escape script*, individuals wish to avoid the discomfort aroused by unpleasant situations, conflicts, or memories. Failure, rejection, disappointment, hurt, humiliation, embarrassment, discontent, or sadness all tend to demand relief. Rather than openly confront a critical employer or unappreciative spouse, endure boredom and frustration, or lie awake at night worrying, individuals only want to eliminate their misery and to quell their harassing thoughts. They are tired of feeling hassled, lousy, and upset. They just want to get away from it all and, more to the point, from themselves. It is not necessarily intoxication they want. They only want numbness, the absence of problems, and peace.

Yvonne portrayed this type of thinking well, shortly after the breakup with her boyfriend. She had known the rift was coming, but still it was a shock. For the first couple of days after it happened, she numbly went about her business, forcing herself to cope, but then difficulty with sleeping started and one morning she awoke just wanting to scream. On that particular morning, she finished in the bathroom, put a kettle of water on the stove, and then wandered into the pantry, aimlessly searching about. But for what?

"God, I'm pissed," she muttered silently to herself. "After all I did for the bastard. You can't trust any man. They're all shits. How could I be so dumb . . . so naive . . . so idiotic? It's just not worth it, staying sober. For what? Why do all these things happen to me? I sure made a mess out of my life. I'm tired of all this crap. I just want out. Away from these fucking thoughts. Oblivion. Peace."

She stood on the stool and, apparently without awareness, reached toward the back of the shelf for the bottle of gin that had been stashed away there some time ago.

"I need a martini . . . something," a voice inside of her declared. "Something powerful . . . strong and fast. I need to stop thinking about this . . . and go blotto. It's driving me crazy. I need to stop this pain, this terrible pain of living. So what if I take a drink, maybe two. Nobody really cares anyway. I need relief, and this is the only way to get it. I just want to be able to laugh again . . . to have a good time."

The craving, which had been absent for years, now began to consume her. It was like being in a room, and on every wall there was a movie being projected, representing a part of her life. Everywhere she looked she saw things she didn't want to see, all the times she had been hurt and disappointed. She wanted to shut out those memories and experiences and not be afraid, not feel like a failure, not feel that awful sense of hopelessness, despair, and loneliness. Her hands trembled as she found the drink that she had no memory of preparing. As she brought the glass to her mouth, all the images began to fade. All the terrible feelings ceased to exist, and it was just she and the booze.

The *relaxation script* also takes many forms. Thoughts of wanting to unwind are perfectly normal, but where they go awry is when they are coupled with expectations of this happening instantly, without benefit of doing something relaxing. Instead of equating relaxation with a walk on the beach, athletics, games, a cookout, a massage, a ride in the country, listening to music, reading a book, or a host of other emotionally restorative activities usually associated with fun, diversion, or leisure, individuals begin to prefer the more immediate and profound experience that can be induced chemically.

In Richard's case, he had started out his vacation with ordinary expectations for relaxation. It had been a hard year for him, with a full-time job in an engineering firm, night school, and AA meetings two or three times a week. He could spend only snatches of time with his family and was still trying to get his life back in order after the separation and later reconciliation with his wife. So his vacation with the family was especially welcome and well-deserved. He had been looking forward to the leisurely drive up north in the house trailer. But after a day on the road, with two cranky boys and his wife snapping at them, he became more and more irritable. He kept telling himself to relax, to have a good time, but the uneasiness and tension continued to grow.

By the time they finally pulled into the state park, after another two days of travel, he unaccountably felt ready to explode. Because he was sweaty and tired, he felt a swim would help. They found a parking spot and then hurried down to the lake. The sun was hot. The water was cold. But he did not feel refreshed after a swim. He got out of the water, dried off, and became aware of his thirst. "It's almost five o'clock, time to relax," he thought automatically, then saw the image in his mind of several beers nestled in a big chest filled with ice. "God, what I wouldn't do for a swig. Just to quench my thirst and relax. Unwind after the long drive. A reward for working so hard." And along with these thoughts, an image popped into mind, an image of himself sitting there drinking beer after beer, like in the old days, feeling totally mellow and relaxed and experiencing a wonderful sense of contentment. He could almost taste the cold beer washing down his throat and then feel the warm glow. Fighting off these thoughts, he found himself wondering whether they had enough provisions for supper, whether he should offer to drive back alone to the nearest store . . . the store where he had noticed the beer signs in the window when they drove by.

The *socialization script* overlaps with that for relaxation but is limited to social situations. Many individuals, shy or uncomfortable at parties or gatherings, may experience the need for a "social lubricant" to feel more at ease and decrease the awkwardness and inhibitions they feel around others. They would like to seem charming, witty, or attractive, like others they encounter, but they just don't know how. Extremely self-conscious, they worry that others are aware of their inadequacies and, therefore, hold them in low regard. These thoughts are bad enough, but they are compounded by their own anger toward themselves for being so concerned about what others think.

Marvin was that kind of person, always worrying about what others thought, especially in social situations, even though, to others, he seemed self-assured and at ease. Unfortunately, being a businessman, he had to interact with the public, entertain prospective clients, and attend all sorts of civic functions—activities that made him very uncomfortable. For the first several years after he quit drinking, he tried to maintain a constructive attitude, reminding himself of how necessary these interactions were for his business and how unproductive it was to dwell on his negative reactions to something that could not be avoided anyway.

But then, after more than four years of abstinence and everything seemingly going well, his attitude about these functions began to change for the worse, and it became harder to psych himself up beforehand. Instead of the usual annoyance, he felt a growing sense of dread that sometimes led to headaches or forced him to search for excuses not to go. "It makes no sense," he told himself, even as he became more self-conscious and socially insecure. Then, when he did attend certain social functions, like parties, he would berate himself for being so awkward and shy. "Others are enjoying themselves, so why can't I?" he would wonder. "Why can't I let go and just have fun . . . not be so self-conscious . . . not give a damn what anyone thinks? Just cut loose for once, say what I want? Why am I so stupid? They must take me for an idiot." Then, to compound his discomfiture, he would convince himself that he was ruining the evening for his wife. "Anne loves parties," he would think. "She enjoys people. She really needs to get out more. And here she is married to such a tight-ass."

But lurking in the back of his mind all these months was a solution he didn't want to think about—the knowledge that with a few drinks, all his concerns would be resolved and his inhibitions gone, as in the past, before he quit drinking.

The *improved self-image script* is similar to the socialization script, but the element of low self-esteem tends to be much more pervasive, extending even to times when alcoholics are by themselves. When individuals start becoming unhappy with themselves, when they are feeling inferior to others, when they regard themselves as lacking in essential qualities, when they feel unattractive or deficient—especially when they have found no other way to alter these feelings—it is only natural that sooner or later they should begin once again to think about that magical potion which, in the past, immediately could remedy all these imagined lacks—at least, temporarily. With alcohol, they have no need to ply themselves with ineffective arguments about their self-worth, to constantly try to prove themselves to others, or to always seek approval or acceptance. With alcohol, they possess a cheap, easy, and immediate solution to what might otherwise take a lifetime to accomplish—feeling good about themselves.

That situation applied to Hastings. By any reasonable standards, Hastings should have had a brilliant academic career. He had a facile mind, a penchant for languages, a gift of expression, and a command of the literature that was the envy of his colleagues. For years, he had been working on a definitive biography of a well-known scientist, a three-volume work that would justify the great expectations everyone had of him and establish him as the world's leading authority on that subject. But the criticism of his manuscript had been devastating, and he was never able to get it published.

The disappointment he experienced over this failure created a bitterness and frustration that was to haunt him the rest of his life. All he wanted to do was hide. His credibility as a scholar was ruined; he had been exposed as an academic fraud. And he was sure that all his colleagues and students knew.

In the past, before this happened, he used to enjoy an occasional scotch, but now he started to drink excessively on occasion, either drinking until he became physically ill or did something embarrassing. Then, repentent and disgusted with himself, he would remain abstinent for long stretches of time, until he started drinking all over again. Typically, days or weeks before a new binge, seemingly inconsequential events would make him depressed and liable to judge himself harshly. He would learn that a colleague had had an article published, and that would stimulate him to think, "Why couldn't that have been me?" Then, feeling a twinge of guilt over his envy, he would answer his query with, "Because, you stupid oaf, you just don't have what it takes." Or he would read a glowing review of a book in a prestigious journal and think, "I'm a failure. It's too late. My career is over." Or he could receive an unenthusiastic response to a lecture and think, "They can see through me. They're not interested in what I have to say. I'm a terrible teacher." Or the department chairman could inquire about what he was currently working on, and, as he defensively muttered something or other, Hastings would think, "He knows that

I've given up on trying to get published. As far as he's concerned, I'm deadwood. He's only asking that to embarrass me."

Then, when Hastings would get this way, he would sit alone in his office or in his armchair at home at night and imagine how it all could have been different. He would imagine his work being published and receiving critical acclaim. He would imagine himself holding court with a coterie of graduate students who were hanging on his every word. He would imagine impressing his colleagues with his brilliance and entertaining them with his wit. He would imagine being offered a prestigious editorial appointment or even becoming chairman. And with just a few drinks, these dreams could become ever more real, and then with just a few more, it wouldn't matter anyway.

With the *romance script*, individuals indulge in adolescent fantasies or Hollywood dreams. Bored or unhappy with their lives, they yearn for excitement, romance, the joy of flirtation and the thrill of being in love. This is usually the kind of thought that, when engaged in too seriously, requires a drug like alcohol to sustain it and make it more vivid and real.

Melanie was a hopeless romantic. After two failed marriages—each resulting in her being hospitalized for drinking—and then trying to raise two small children on her own while working full-time, she should have had no illusions left. To her friends she admitted that she had had it with men, yet when she was alone at home or bored, she would think of some new man she met and how different it could be. Not all men were alcoholics, like her ex-husbands, or like all her eligible male friends she had met at AA. God, she was sick of being around alcoholic men, starting with her father and an uncle. "I just want to meet someone normal," she would think to herself. "Then maybe it could be different. Just go out on an ordinary date. He doesn't have to be handsome, but he has to have a sense of humor and enjoy dancing. I don't know how long it's been since I've done something like that." Then, in her fantasies, she would picture herself sitting in a dark corner of a nice restaurant, engaging in intimate conversation. She would almost feel her heart tripping during the magic moment when her date started to pour the wine. Sometimes as she played out these fantasies, tears would come to her eyes.

With the *sensual pleasure script*, individuals mainly think about their own physical gratifications. The thought content and mental imagery is hedonistic in nature, oriented toward self-indulgence and more immediate pleasures. A delectable, subtle taste. A fine bouquet. The smooth, silky flow of liquid down one's throat. A warm, caressing sensation, emanating from within. An icy, cold drink on a hot, muggy day. The beautiful, amber color of a fine scotch. The delightful tickling sensation of a bubbling, effervescent liquid within the throat. When individuals permit their minds to dwell on images such as these, their thoughts eventually become filled with pleasant recollections of drinking, often causing them to experience many of the intense cravings that dominated them before. These recollections about the physical delights of alcohol are more likely to promote its actual consumption than some of the more abstract thought processes associated with most of the other mental scripts. With these other mental scripts, craving is apt to come, but only much later.

In Joe's case, this kind of thinking led to his downfall. Joe had been a sensualist of sorts, but he preferred to believe that phase of his life was over. Now, he was just trying to stay dry and make something of his life.

Joe liked his construction job well enough, and he didn't even mind the ribbing he got from the guys about being henpecked, although he did think they probably were right. The reason he went home right after work and didn't stop at the local bar with them as they kept inviting him to do—-and this was something they didn't know—was because the drinking had got him in trouble before on his last job and almost caused a divorce. Still and all, he thought his wife was being unreasonable, worrying whenever he didn't show up right on time. She had to learn to trust him. He had said he was through with drinking, and she should believe him. After all, it had been almost a year since he last had a drink. That proved something. And there would be nothing wrong, just to show he wasn't snooty or stuck-up, if every once in a while he joined the fellows after work and had a soft drink or two just for social purposes.

So he tested himself one week, and sure enough he was right. A few sodas, and he had no temptation for anything else, even though the guys did tease him about not drinking. It was possible to enjoy himself without alcohol. But his wife disapproved. At first she expressed concern about his coming home late again, then warned him against drinking, then carped about him neglecting his family, and finally resorted to the silent, martyr treatment, which always infuriated him. "Christ," he thought, "I might as well stay out and get drunk for all the credit she gives me. I get the same response anyhow, whether I drink or don't. But I'm not going to let that throw me."

And then one very hot, muggy day after work, when he was feeling thirsty and irritable, he took one sip of the coke and found himself sickened by it. "Ugh, it's sweet," he muttered to himself, and watched the other fellows happily guzzling their beer. But for some reason, this time the desire to join the others in a beer was compelling. "Not that I crave it," he told himself, "but just for taste and to be social. Besides, it wasn't the beer that got me in trouble last time, it was the booze. I always could handle beer." Then he remembered other times, like at a barbecue, and how great the beer tasted as he washed down the food. At that moment, if he could have translated the images and sensations he experienced into words, he would be saying in his mind, "There's nothing like the malty flavor of beer. I can just see the beer being poured into a frosted mug, looking like it's really cold, with little droplets of condensation on the glass. I can feel the coldness and wetness going down my gullet and then the warmth and fullness spreading in my belly. It's not so much the effect, the buzz, I'm after, but it's that delicious taste and being able to quench my thirst. There's nothing like it on a hot day. I don't need it, but I sure want it."

These images and thoughts could be resisted for an evening or two or even a week or more, but sooner or later, if they remained or were nursed, as Joe later would readily admit, they would give way to a monumental thirst for beer that would become increasingly difficult to curb.

With the *to-hell-with-it script*, individuals seem to have lost all incentive for pursuing any worthwhile goal. Their thoughts express disillusionment. Nothing really matters. There is no reason to try. Nobody appreciates the effort anyway. Why should they give a damn? So these individuals let their mental guards down and become less vigilant against the forces that may lead them astray. So what if they are being self-destructive? What difference does it really make anyway? And so they reason, often realizing full well where this kind of thinking will take them. There are no brakes, no restraints anymore for their impulses.

This was the frame of mind that eventually contributed to Dr. Sam's relapse. Ever since he had made the decision to join AA over five years ago, Dr. Sam, as he was known to his patients, continued to build a thriving dental practice. But over time he derived less and less satisfaction from that. Going to the office each morning became like pulling teeth, figuratively and literally, a procedure he hated even more than filling cavities. Several times he felt like he was on a treadmill, getting nowhere but unable to get off. "What does it matter?" he soon began to think. "Nobody really appreciates what I'm doing anyway. All Alice cares about is the house, the kids, and the financial security. She doesn't really care about me. I'm tired of trying to please her, of taking all that crap from patients, or working all those extra hours, just so she can buy those fancy clothes or send money to her relatives. And I'm also getting disgusted with the kids. All they seem to care about is handouts. Hell, they'd rather be with their friends than spend any time with me. They even seem ashamed to be seen in public with me. So why am I busting my ass? It's just not worth it. To hell with it. Why shouldn't I be able to goof off and have some pleasure too? Everybody is only out for themselves anyway, so why not me? What difference does it make? I don't care anymore what anybody thinks. From now on I'm just interested in number one."

With the *self-control script*, individuals are bent on proving to themselves that they are not alcoholics, that they can, in fact, regulate their drinking. When it suits their convenience, they suffer from selective memory loss, forgetting all the times they were unable to quit drinking when they wanted to, the many times they drank much more than they intended to or should, and the many times that they told themselves the same thing as now only to be proven wrong. Characteristic of their thinking is a false sense of confidence, a feeling that they are once again in full control. All that is necessary is to make up their minds, to decide beforehand just how many drinks they can have and then to stick to it. Now that they know what can go wrong, they know how to prevent it—or so they believe.

Boris followed this script. He was bored going to AA meetings. All that mumbo jumbo about a Higher Power and all those sickening, drunkalogue confessions. He was tired of listening to all that junk. He didn't know why he let his boss talk him into going in the first place.

"I'm not an alcoholic," he insisted to Karen, his girlfriend. "I can take it or leave it. A few bad hangovers and missing a few days work doesn't mean I've got a problem. Hell, you should listen to those guys in AA. They're real drunks, d.t.'s, jail, losing their jobs . . . the works. Can't control their drinking. One drink and they've had it. It was never like that with me. Oh, sure, I've tied a few on in my day, but nothing like what some others do. But then, again, I've gone for days without a drink and never missed it. Just because someone gets high a few times doesn't make him an alcoholic. The important thing is to know how much you can handle. There are some who can and some who can't, and I'm one of those who can. Just going through all this nonsense, I've learned a lot. The key is to set a limit beforehand about how much you should drink or to stop just as soon as you feel the first signs of being high. Don't keep drinking until you're drunk. Heck, being drunk isn't any fun anyway. That's where I've made my mistakes before, drinking more than I should. It's just a matter of planning ahead, self-control and being intelligent about what you do. The next time it will be different. You've got to use willpower."

The *no control script* represents the other side of the coin. Just as believing in one's ability to handle alcohol intake is usually a setup for relapse, the opposite attitude of not being able to control one's cravings virtually insures it. Individuals give up the fight, conceding defeat even before they have made any effort to resist the urge. This attitude is not limited just to alcohol. It pervades almost all aspects of their lives. They feel that they simply have no power to better relations within their families, to improve their economic prospects, or to change their general situations. They feel powerless, impotent, and defeated. The automatic retreat to alcohol becomes a natural elaboration of this outlook. This attitude, it should be emphasized, is not the same as to-hell-with-it. With that kind of script, individuals do not necessarily feel powerless. They just do not want to exert the effort to continue what they have been doing.

Vince illustrated the no control scenario well. In his case, there was no question about what was bothering him. His life was a mess, and his self-confidence was at an all-time low. It had been over six weeks since he had been discharged from the alcohol-treatment unit at the Veterans' Administration hospital and everything was worse than before. He figured that by volunteering for treatment everything would improve, but a month in the program did not change the fact that he had been fired from his job, that his wife had left him, that he soon would have to go to court, and that he still was spending most of his waking hours fighting off the desire to drink.

"Why try to fight it?" a voice in his head would argue. "You're an alky. It's hopeless. You can only hold out so long. Why torture yourself? You know you don't have any control."

Then another voice would answer, "Don't listen to that. You've held out this long, you can hold out longer. You're just feeling sorry for yourself and looking for an excuse."

Then the first voice, the persistent and convincing one, would counter, "Stop kidding yourself. It's just a matter of time. So why torture yourself? You can try to hold out, maybe go to an AA meeting or call someone, but that's only putting the finger in the dike. You're a miserable son-of-a-bitch, a weak-willed bastard, and everyone knows it, so why don't you wise up? It's no use. It's not worth the effort.

You're totally helpless. You've got no control. So give in already. It's going to happen anyway, sooner or later, and it might as well be now. No sense in holding out. Go ahead, take that drink."

Though probably not exhaustive, the mental scripts described are typical of what transpires in the minds of alcoholics preceding relapse. Despite their diversity, these scripts all share one feature in common. They reflect a basic discontent or unhappiness with the way things are imagined to be, have been or will be. It is difficult enough for recovering alcoholics to stay dry, but when they constantly believe that something is wrong or amiss in their lives, then they continue to remain susceptible to the instant, easy solutions that alcohol has to offer. This state of mind provides a fertile soil for craving.

4

The Mystery
of Craving

A fundamental issue that needs to be addressed about alcoholism is why otherwise reasonably intelligent and competent men and women should find it so difficult to resist the temptation to drink and, once they have surrrendered to it, should be unable to stop consuming alcohol before reaching the state of intoxication or stupor. What is this seemingly irresistible urge which eventually gives way to an unquenchable thirst, even in individuals who struggle to remain sober? For those who have been dominated for years by this urge, an equally important issue is how it can stop, as it apparently does for some, once they decide to quit drinking. How can so insistent and persistent an urge suddenly disappear?

My personal experience with another addicting substance, nicotine, gave me a firsthand experience with these issues. For more than twenty-five years, I had been an inveterate smoker, first of cigarettes, then a pipe. In these latter years, every occasion, no matter how minor, served as an excuse for me to fill, tamp down, and then light up another bowl of tobacco. A cup of coffee after breakfast, the drive to work, a morning conference, answering mail, treating a patient, lunch, a walk outdoors, a coffee break with colleagues, arduous mental activity, a martini before supper, television after supper, reading a book, a social event—all served as occasions for me to seek out the distinctive taste of tobacco smoke, nursed lovingly for a long moment in my mouth after each precious inhalation and regretful exhalation. This despite the rawness of my hard palate by the end of the evening, patchy ulcerations on my tongue, tobacco-stained teeth, a chronic sinus condition and cough, and the certitude that I would be developing cancer of the tongue, throat, or perhaps even lungs if I continued to smoke. Every night I vowed to stop this insane habit. Every morning I gave in to my intense craving for tobacco.

Then, toward the end of February 1980, I developed a bad sore throat. I still tried to smoke my pipe, but it was just too painful so I decided to stop for a few days—as I had done many times before under similar conditions. As my throat began to improve, I found myself looking forward to the luxurious smoke that would soon be circulating about the cavern of my mouth. But then a contrary desire seemed to possess me, and I suddenly made up my mind to quit. This was it. I had not smoked for three days and managed to survive. It was insane for me to start again. I knew with certainty, unlike the many times I had tried to quit before, that this time I would be successful. I told myself that tobacco no longer had any hold over me; I was rid of my addiction. From that moment on, the craving for tobacco was gone, and I never again experienced any conscious desire to smoke.

The question is: What happened to the craving? No natural, scientific explanation seemed adequate to account for the sudden disappearance of this almost incessant, obsessional urge that had dominated my daily life for so many years. How could so overpowering an urge vanish so completely? Was it lying dormant, awaiting to be reawakened by my first puff of smoke, or was it gone forever, exorcized like a demon? I sensed it was still lurking within some deep recess of my mind, but I couldn't be sure. And I didn't want to try to find out.

My experience stood in stark contrast to that of my wife, who quit smoking one month later. For the first year or two, she acutely missed cigarettes and likened the feeling that their loss gave her to "the death of a friend." The craving was with her almost constantly, even during sleep. By the third year, she still had powerful urges to smoke, particularly in the evenings, but the general frequency and intensity of the desire mercifully began to abate and then, finally, disappear as she struggled her way to success.

These accounts are not so different, as we shall see, from those described by many recovered alcoholics. After years of "hopeless" addiction, they report that their craving vanished suddenly, as if by divine intervention, as had happened with me, or it lingered on to torment them for years after quitting drinking, sometimes like a succubus, as happened with my wife, even intruding into dreams. "After 13 years," John O'Hara wrote, "I can still taste Scotch and beer, and I still have a recurrent dream about Scotch, in which I am at the bar at 21, order a St. James, am just reaching for it, and Emil takes it away. I have another dream in which I belong to three non-existent clubs, two in New York and one in Philadelphia, to which I go for a sneak drink, but when I get to the clubs they are all closed."[1]

It is interesting that so insistent and persistent an experience as craving should itself be the subject of scientific controversy, not only about its nature but its very existence. Strict behavior-oriented scientists, who place more value in observable actions than in what takes place in the human mind, tend to regard craving as a logical tautology, a completely unnecessary notion, because it is supposedly defined by subsequent drinking behavior.[2] The argument goes that if every time someone consumes liquor he says he does so because he craves it, then why bother to be concerned about craving? It explains nothing and complicates matters. It may not even be "real." The appropriate object of scientific concern should be the drinking itself, which can be observed and directly measured, not the subjective experience of craving, which cannot.

Various experimental studies and reports are usually cited in support of this view.[3] Alcoholics, for example, do not report an increase in craving if given a drink with the taste of alcohol disguised. When they don't know they are drinking alcohol, they don't necessarily crave it. In laboratory or ward settings, alcoholics seldom drink to oblivion: the amount of alcohol consumed usually proves to be inversely related to the amount of work necessary to obtain it. The more they have to work for alcohol, the less they drink. Not surprisingly, like many other individuals, alcoholics can also be bought. If paid enough money, they are willing to refrain from drinking. It is simply a matter of finding the right price. What these studies supposedly show is that craving, which can be manipulated by different means, does not seem to be an important factor in why alcoholics drink. So why give the notion any credence?

The behaviorists make a convincing argument, if true. The error is with their assumptions. Just as hunger need not always be expressed in eating, craving need not inevitably be associated with drinking, so no tautology exists. Sometimes individuals can give in to this urge, other times not. Sometimes they feel compelled to drink until drunk, other times not. And sometimes they drink without experiencing any urge to do so at all. Why individuals respond differently at different times represents the heart of the issue.

Two accounts illustrate these different expressions of craving. Don Musgraves, the author of *One More Time*,[4] had managed successfully for some time to keep his desire to drink in check, until one day when he found himself wrestling with the decision of whether to steal a large sum of money from his employer. At an AA meeting earlier that day, he had been restless and short-tempered. When he got in the car, he had trouble driving, so he pulled into a nearby parking lot and sat for a while, trembling all over. On his tongue, in his throat, burning its way down his esophagus, he could feel and taste something he hadn't even smelled for nearly a year-and-a-half—whiskey. All of a sudden, he had a wild, consuming thirst.

He realized that if he had one drink—just one—his eighteen months of sobriety would go down the drain. "What the hell, why don't you just rip off that dough and split?" a voice roared in his brain as an icy sensation traveled down the back of his neck, into his legs. Images of his wife and kids prompted him to think, "You have lost your marbles." He prayed and tried to remember some of the words about God in *The Twelve Steps* but they wouldn't come. The only words he could groan out were, "Okay, you win, stop tormenting me." Then he drove to the store and stole over $5,000 from the registers and safe.

All that time he had a sickness in his gut, as though something still sane inside was holding on for dear life but was powerless to stop him. From there he went straight to a bar, ordered a drink, and for three hours sat staring at the glass. He breathed heavily, struggling with himself, but in the end, suddenly lifted his glass and gulped down the contents, unable any longer to resist the urge. Then he left the bar and ran off with the money.

This account stands in stark contrast to that of a physician I interviewed who was equally tempted to drink. The episode took place on Saturday afternoon, after a three-year period of abstinence. That morning, the doctor had gone to his customary AA meeting, but this time he became unaccountably impatient and angry and walked out before it was over. When he got home, instead of going to the university basketball game as planned, he gave his tickets to his kids instead. With his family gone, he planned to watch the game on T.V. He ostensibly preferred to watch the game alone, without distractions, because he could follow the action better, or so he rationalized. Still, he found himself mad, not knowing why. Then all of a sudden he started thinking about a drink. It was 12:30 PM. The game was scheduled to start at 1:00 PM. Liquor was in the house, mainly for guests, but he had hardly thought about it in the past. During this thirty-minute period, he went "absolutely bananas." He kept telling himself over and over again, "I can't drink, I'm an alcoholic." But then a voice inside countered with, "Yes, I want a drink, I want to get comfortable, to feel that old feeling." He was shaken by the powerful pull of these compulsive thoughts and the sense that he was about to lose control. In about ten minutes, he called a friend in AA, but interestingly enough, talked about someone else, another physician, never mentioning what was going on inside him. He talked about ten minutes, hung up the phone, and then went back to the same thoughts. He had to have a drink. The ballgame was going to start any minute. He went to the refrigerator, grabbed some food, and began eating to shift his attention. He read some AA literature, hoping that would help. By now, the ballgame had started. For the first half of the game, he didn't remember anything except constantly thinking about having a drink. He even got out of the chair once, saying out loud, "I'm just going to have a little taste," but then sat right back down. The temptation persisted. He decided to call somebody in AA again, but then talked himself out of it. The compulsive thoughts continued. Finally, he thought of the Serenity Prayer. "God grant me the serenity to accept the things I cannot change," he said aloud, finishing the prayer that he knew so well. At that very moment, the craving left and he never experienced it again.

What these anecdotes demonstrate is that it is quite possible for someone to experience a consuming thirst for alcohol and either act on it or not—that craving is not necessarily equivalent with the consumption of alcohol—and that the final capitulation to the urge or mastery over it can be influenced by other factors. In Don Musgraves' case, his struggle over whether to steal apparently released a craving that had long been dormant, but it was only acted upon after he had made the decision to keep the money and run. In the physician's case, his sudden urge to drink also seemed to stem from his prior state of mind—the unexplained anger and annoyance he felt earlier in the day and his "unwittingly" setting himself up for a drunk by sending his family off to the game—but some inner reserve and resolve helped him to resist.

Unfortunately, it is far simpler to demonstrate the importance of craving as a determinant of alcoholic behavior than to define just what it is, especially when the concept has been used in so many different ways. An expert committee of the World Health Organization[5] pointed out that the concept of craving has been invoked to describe such different conditions as the onset of excessive alcohol use, the drinking behavior shown within a drinking bout, relapse into a new drinking bout after days or weeks of abstinence, continuous daily excessive drinking, and loss of control over drinking. The role of craving is also a matter of debate. Craving, as well as its equivalent or alternate terms— urge, need, appetite, desire—have been used to explain drinking as arising from a psychological need, from the physical need to relieve alcohol withdrawal symptoms, such as a hangover or the shakes, or from a pathological, metabolic need that either exists in the drinker before he starts on his drinking career or develops during the course of it.

In order to reduce confusion, more precise definitions are needed. According to Dr. Harris Isbell,[6] former Director of the Addiction Research Center at what was then the U.S. Public Health Service Hospital in Lexington, Kentucky, there are two types of craving. "Nonsymbolic" craving occurs in alcoholics who have drunk excessive amounts of alcohol over a protracted period of time. It expresses itself through the appearance of symptoms upon withdrawal of alcohol— symptoms like agitation, shakiness, headaches, confusion, hallucinations, or seizures—and is believed to be due to physiological alterations, the mechanisms of which are not completely understood. Though physiologically determined, these symptoms may be colored or altered by psychological factors. In contrast, "symbolic" craving, the type which warrants our special interest, is presumably psychological in origin, accounting for the initial misuse of alcohol after a suitable period of abstinence.[7] While the distinctions between these two types of craving at first seem clear—nonsymbolic craving representing a physical need for alcohol during or immediately after a heavy drinking bout, and symbolic craving representing the intense desire for alcohol during a period of sobriety or preceding a drinking bout—difficulties arise when one tries to determine when the physiologically based craving ends and the psychologically based craving begins, particularly when it has been established that subtle signs of alcohol withdrawal can be detected for at least six months or more after the last drink. Does this transition occur after one week following withdrawal, one month, three months, or one year? Or is the distinction between these two types of craving artificial? It seems more reasonable to assume that nonsymbolic craving for alcohol can never be divorced entirely from psychological factors, and symbolic craving can never be divorced entirely from physiological factors. The proportion of these factors would be a matter of degree, dependent upon such variables as the length of time the individual had been drinking, the amount of alcohol consumed, and the relative interval of time elapsed since the last drink, as well as the individual's expectations and the circumstances at the time.

Another feature of craving, whether of physiological or psychological origin, is that it tends to vary in strength.[8] A sense of tension, unease, irritability, restlessness, insecurity, self-disgust, vague guilt, or annoyance are all associated with mild craving; distress, depression, shakiness, apprehension, jumpiness, a knot in the stomach, fear of dying, a feeling of sickness, or being out of control, are associated with strong. This means that craving is not an either-or phenomenon. The stronger it is, the harder it is to resist.

It is understandable that those who discount the importance of craving should wish to avoid traveling in the murky realm of subjective experience, which tends to be inhabited by intangible and ineffable drives, urges, and passions, but the difficulty in observing and measuring craving does not make it any less real. There are times, in fact, when it is almost as palpable as any material object. Consider, for example, the personal recollections of former Senator Harold E. Hughes.[9]

"I had picked up a copy of the evening newspaper and was scanning the pages when I suddenly felt the urge. By that time, I had not touched alcohol for over a year, and though there had been many urges, I had been able to overcome them. However, my longtime habit of an evening drink coupled with being alone in a hotel room generated a powerful force deep within me. I wanted a drink. I needed it. I had to have it.

"Desperately I battled. I turned back to the newspaper and tried to read. Drumming incessantly within me, however, was the demand for a drink.

"I stood up, the paper falling to the floor. Suddenly, I felt like two different people, the new and the old Harold Hughes. The urge became overpowering. I knew that in a very few moments I would be going to dinner at a downtown restaurant. To reach it I would pass an old drinking spot. And I knew as well as I stood in that hotel room that I would go into my old haunt for a drink. I could already savor its delicious burning strength."

Or, if that is not a dramatic enough portrayal of the power of craving, consider the experience of someone I knew who, though sober for some time, was still battling the urge. His desire for alcohol was so great that at times, in desperation, he would try meditation, yoga, Zen, arguing with himself, praying on his knees, eating honey, taking a hot bath, reading, smoking—in short anything and everything to keep busy and distract himself until the urge passed. For both Senator Hughes and this man, the craving for alcohol may have been intangible and invisible, but when it hit them, it became one of the most overpowering realities in their lives.

What complicates an understanding of craving is the extent to which this experience can be influenced or molded by psychological factors and expectations. How can something be real, so skeptics argue, when it depends so much on how someone thinks? But this is just the point. Craving is a subjective experience, so it should not be too surprising that it can be affected by what goes on in the mind. The sexual urge can be influenced by psychological factors. Hunger can. So why not craving? It is this very susceptibility to manipulation by psychological expectations that constitutes one of its most characteristic features and probably represents the very "nature of the beast."

Perhaps the clearest illustration of the impact of psychological factors or the power of suggestion on the drug experience involves a study with narcotic addicts[10] at the former U.S. Public Health Hospital in Lexington, Kentucky, in the early 1960s. This particular investigation, among others, was designed to determine the relative importance of psychological and pharmacological factors on the actual drug experience. While under deep hypnosis, the participating addicts were told that they were going to receive a saline solution intravenously, when in fact they received the narcotic drug, morphine, or that they were going to receive morphine intravenously when in fact they received salt water. The results were dramatic. Those who were told they received morphine (but actually had received salt water) yawned, scratched, acted euphoric, and eventually went "on the nod"—all characteristic effects of the narcotic—while those who were told they received the salt water (but actually received morphine) showed no unusual behavior even though significant physiological changes in pupil size, heart rate, and respiration took place. These findings imply that if addicts, under certain conditions, can get "high" on salt water or not get "high" on morphine, then alcoholics, under comparable conditions, should also be profoundly influenced when they are drinking by their notions of what to expect.[11]

The "think-drink" effect—a term referring to the experience of craving when alcoholics think they are consuming alcohol, even though they may not be, and the lack of craving when they don't believe they are consuming it, even though they may be—makes just this point. This effect was shown in a well-known study[12] in which both alcoholics and social drinkers were given either tonic water alone or a vodka and tonic without their being told exactly which they received. It was found that those who thought they were drinking vodka and tonic drank significantly more than those who expected only tonic water, regardless of whether there was any alcohol in the drinks. This indicated that the expectations of alcoholics played a far more important role in determining how much they drank than the actual consumption of alcohol.

In related studies[13], my colleagues and I have conclusively demonstrated that environmental cues can exert a profound effect on expectations. In barroom settings, for example, alcoholics experience far more craving than in environments not conducive to drinking, such as a hospital laboratory. In other words, alcoholics are apt to crave alcohol and drink more in situations that allow them to believe it is alright for them to crave alcohol and drink more.[14] This clearly indicates that if recovering alcoholics want to minimize the temptation to drink, they should avoid settings in which drinking is allowed or encouraged and seek out "safe havens," like workplaces, their church, clinics, and AA meetings, where it is not.

But that is not all of the story. The fact that attitudes and expectations can have a profound impact on craving does not negate the po-

tential effects of alcohol itself on craving. Under the proper circumstances, like in settings conducive to drinking, alcoholics tend to
experience craving to a much greater degree when alcohol is added to
a sweetened soft drink than when it is not. Interestingly, one to two
ounces of 100 proof alcohol tends to produce much more craving than
a much larger amount, like six or more ounces. Like the hors d'oeuvres
before a main dish, the smaller amount of alcohol seems to whet the
appetite for more, a finding that supports the wisdom of the Alcoholics
Anonymous injunction to "avoid the first drink."

Small amounts of alcohol are not the only chemical triggers for
craving. There is evidence to suggest that almost any drug that produces effects resembling those of alcohol may stimulate the desire for
alcohol, provided, of course, that it is taken under the "proper" circumstances—namely, conditions that promote drinking. For example,
low amounts of tetrahydrocannabinol, the active ingredient in marijuana, which may have subjective effects similar to those of alcohol,
can also produce a strong craving for alcohol.[15] These pharmacological cues are probably the reason why certain drugs that alter the
mind, like sedatives (e.g., barbiturates) or minor tranquilizers (e.g.,
Librium, Valium), which have been widely used to treat anxiety in
alcoholics, may actually serve to keep the embers of craving alive,
continually reminding otherwise abstinent alcoholics what it feels like
to be mildly intoxicated. If they do not relapse while taking these drugs,
then they may find themselves simply substituting a dependency on
alcohol for a dependency on drugs with alcohol-like effects.

This still does not explain the nature of craving or the function it
serves. A better understanding can be gotten from what is already known
about an equally compelling drive—the "hunger" for narcotics. That
is where most of the pioneer work on addiction has been done.

The research began over thirty years ago with an incidental observation made by Dr. Abraham Wikler, a psychiatrist and neurologist,
who was then working at the Addiction Research Center of the USPHS
Hospital in Lexington, Kentucky.[16] Dr. Wikler was puzzled by the
frequent reports of many treated heroin addicts who left the hospital
feeling confident and hopeful but who, on the train back to New York

or in their neighborhoods, suddenly became sick. Usually, they regarded their sickness as the flu, but the symptoms they experienced—muscle aches and pains, runny nose, watery eyes, sweatiness—were strongly suggestive of early narcotic withdrawal symptoms. Because of this tendency to misconstrue what was happening to them, they usually began to have a yen for "God's own medicine"—heroin—which had worked so well to "cure" them in the past. As before, the heroin produced dramatic results, but what was to have been one "fix," led, much to their disgust, to many more "fixes" thereafter, and before they realized it, they were "hooked" again.

These observations suggested to Dr. Wikler that these symptoms of sickness common to exaddicts trying to stay straight were really conditioned responses to certain environmental cues associated with past drug use.[17] The hunger for the drug, he proposed, arose for the same reason that the sight, smell, or even thought of a favorite food can elicit salivation in someone hungry or even recently fed. And since drug use tends to occur after the effects of the last dose of drugs has worn off or is wearing off, a time when the drug hunger is greatest and when addicts are most likely to experience withdrawal, it should not be surprising to find that settings or situations reminiscent of past drug use should trigger similar conditioned withdrawal responses mimicking the flu in anticipation of the exquisite pleasure or normalizing effects of a "fix."

The supporting evidence for these conditioning effects is strong. Rats previously addicted to opiates in certain cages repeatedly show signs of drug withdrawal when returned to these cages. A number of studies have also shown that narcotic addicts can be induced to experience conditioned withdrawal symptoms, such as a runny nose, tearing, and general achiness, when exposed to pictures of common drug paraphernalia, like syringes, tourniquet belts, or needles, or of other addicts "cooking up" or "shooting up" a fix.[18]

Extending these observations to alcoholism, we should expect similar patterns to occur. Craving and other minor alcohol withdrawal symptoms, like shakiness or irritability, which have been relieved repeatedly in the past by the intake of alcohol, could be expected to

appear again whenever individuals experience certain feelings or encounter certain environmental stimuli that have been associated with
those prior drinking episodes. There is evidence to indicate that this is
the case. In one relevant study,[19] designed to test the extent to which
alcohol withdrawal symptoms could be brought on by various drinking
cues, sixteen abstinent alcoholics were exposed to two sets of slides,
one alcohol-related and the other nonalcohol-related, on two separate
occasions. The alcohol-related slides included pictures of a state liquor
store, a glass and bottle of beer, a bottle of vodka being poured into a
glass with ice, a bottle of 60 percent pure alcohol along with a bottle
of beer and an empty glass on a kitchen table, a man starting to drink
from a large bottle of wine, and two men sipping on cocktails at a
table. The comparable, nonalcohol-related slides included pictures of
a building with shops; a glass and carton of milk; a teapot and a tea
box; a coffeepot, a carton of milk, and a box of cocoa on a kitchen
table; a man reading a newspaper while seated at a table with a glass
of milk, a carton of milk, and a coffeepot on it; and two men reading
newspapers while seated at a table with a glass of milk, an empty glass,
a carton of milk, and a coffeepot on it. The results were predictable.
In comparison to the nonalcohol-related slides, the alcohol-related slides
produced more significant increases in craving, as well as tension and
anxiety, in alcoholics.

But just as alcohol-related cues, through conditioning, can induce
an unpleasant emotional or physical state prompting relief by alcohol,
they can also trigger pleasant expectations, such as salivation, giddi-
ness, and imagined warmth or even a mild "buzz," in anticipation of
the relaxing or euphoric properties of alcohol.[20] However, the arousal
of craving in anticipation of only pleasure appears to be much less
common in hard-core alcoholics than in those who are not so strongly
addicted. That is because individuals who are more physically dependent on alcohol are also more apt to be drawn to it as a source of relief
from uncomfortable withdrawal symptoms, emotional discomfort, or
as a means of feeling more "normal" than only as a means of getting
intoxicated or high.[21] This is illustrated by the results of one or our
studies, in which between 75 to 93 percent of alcoholics claimed to

experience craving when they were bored, sad, worried, nervous, or depressed, compared to 5 to 35 percent when they were busy, relaxed, or happy. [22]

Another feature about the conditioning of craving needs comment. Many of the cues that elicit craving tend to be highly specific to the individual. Bob Welsh, a pitcher for the Los Angeles Dodgers, for instance, supposedly craved alcohol after golf, after pitching, or when flying in an airplane. [23] Someone I knew had the urge whenever he talked on the phone. Another individual experienced it just before attending weekly therapy sessions with her psychiatrist. And at least several others, interestingly enough, had it whenever listening to the "drunkalogues" at AA meetings.

This individuality of drinking cues is confirmed by the results of one of my studies in which a total of 150 male and female alcoholics were asked to pinpoint the conditions under which their craving appeared. [24] These individuals were told about Pavlov's experiment in which dogs were conditioned to salivate in response to the sound of a bell which had previously preceded the appearance of food. Then they were asked to identify any comparable "bells" that caused an automatic reflex-type craving for alcohol. Over 90 percent of the sample could identify one or more "bells" as triggers for their craving. The wide range and varied kinds of responses were intriguing. With the exception of "inner tension" (including nervousness, shakiness, or irritability), which was the most common trigger for drinking, no other specific cues were reported by a majority of alcoholics, indicating that what may be tempting to some may hardly faze others. The other situations capable of eliciting craving, which were mentioned by up to a third of the alcoholics, included social interactions (such as cocktail parties, dances), meals (which might include a wine list on the table), the end of the workday (for many, martini time), depression, hot weather activities (for example, cookouts and get-togethers), celebrations, music (reminiscent of past drinking experiences), romantic occasions, bars (such as old drinking haunts, barroom atmosphere), pain, alcohol ads, and miscellaneous activities, like flying on airplanes, sports events, sailing, rock concerts, free time, talking to others about drinking, or, in the case of one person, taking a hot bath.

These findings have to be tempered by awareness of another matter: namely, that the number of "bells" capable of evoking craving tends to expand or shrink, depending upon whether individuals are drinking or not and the extent of their drinking. When individuals are drinking heavily and steadily, almost every situation or circumstance may become an excuse for drinking. They drink because they are happy or sad, the weather is too hot or too cold, or they just drink because they drink. When individuals have not been drinking for some time, the range of cues tends to narrow and each becomes more specific. As even more time elapses, the same bells that once clanged in their heads and demanded their attention when drinking eventually come to be silenced or, at worst, become muffled tinkles that can easily be ignored.[25]

What, then, is the function of craving? From what has already been said, it appears that the experience of craving represents an effective way of protecting the alcoholic against any form of physical or mental distress by automatically alerting him to a potential source of relief—namely, alcohol. It also can serve as a directional sign toward pleasure. The individual does not have to say to himself, "How can I feel better?" or "What can I do for fun or relaxation?" Though short-sighted and maladaptive, craving provides the drinker with a clear course of action, an immediate solution to the way he feels. No trial-and-error activity is needed. An alcoholic I knew expressed this matter well. He told me that when he would awaken in the morning and not feel right, he knew he needed something. It could be breakfast, it could be antihistamines, or it could be a walk, but the craving that soon arose indicated that it was alcohol. More often than not, his conclusion proved correct, since he immediately felt better after a few gulps of liquor.

These observations on craving have important implications for the process of recovery. Though many alcoholics may feel helpless and bewildered when craving strikes, they need to recognize that it is not the elusive, mysterious, or uncontrollable force they may think. Far from it. It is a perfectly natural phenomenon, like thirst or hunger, and subject to the same natural laws. To a large extent, its appearance or disappearance is predictable once certain information is known. All individuals have specific cues, some of which are shared by others and some of which are not, that can trigger this urge even without their awareness. They are also susceptible to other cues, such as jail or hospital or church settings, in which alcohol consumption is forbidden, that can dampen this urge. In addition, when craving is reinforced by occasional or continued drinking, it tends to remain at full strength; when it is not reinforced, as during an extended period of abstinence from alcohol, it tends, like any habit, to weaken or disappear over time. It is also clear that craving, when it arises, serves a very important function for alcoholics. It directs them to a potential source of relief or pleasure—alcohol—but, though it may sometimes seem overwhelming, it need not always result in drinking. The fact that craving can exist independently of alcohol consumption—that alcoholics need not act on their urge—is the feature which offers hope of recovery. If the craving for alcohol always led to drinking, then alcoholics would be nothing more than servomechanisms that had to respond reflexly to certain stimuli in certain stereotyped, lock-step ways. But they are not automatons. Through willpower or conscious choice, alcoholics have the option to intervene between thought and deed and alter their patterns of behavior.

5

On and Off the Wagon

Some find it useful to make a distinction between being "dry" and being "sober."[1] Being dry, in this parlance, refers to stopping drinking and eliminating alcohol from one's system but does not mean being committed to this forever. It is not necessarily a voluntary act but can be forced upon the individual by illness, confinement to jail, hospitalization, or the unavailability of alcohol. Being sober, or the state of "sobriety," is something more. It refers to the deliberate substitution of a more positive frame of mind and lifestyle for that associated with drinking. According to this distinction, it is possible to be dry without being sober; it is not possible to be sober without being dry.

Being dry begins with "drying out." For individuals physically dependent on alcohol, but not necessarily for those who are not, cutting down abruptly on the amount of alcohol intake after an extended period of drinking tends to cause a variety of unpleasant experiences, known as withdrawal symptoms, which can be temporarily relieved by the intake of more alcohol. Only a small fraction of alcoholics experience, after a prolonged drinking bout, the more severe and sometimes life-threatening symptoms of alcohol withdrawal known as delirium

tremens or d.t.'s, which may include frightening visions, auditory hallucinations, sensations of bugs crawling on the skin, disorientation,
confusion, agitation, and seizures. The vast majority of alcoholics mostly
experience minor or less serious manifestations, including irritability,
uneasiness, vague fears, fleeting misperceptions, difficulty in thinking,
restless sleep, disturbing dreams, morning sweats, gastrointestinal up-
sets, hangovers, or just feeling "lousy." These symptoms usually appear
within four to eight hours of the last drink, peak at about twenty-four
hours, and sometimes last up to four weeks.[2] Some, who are not yet
so physically dependent on alcohol, experience hardly any physical
discomfort at all.

This drying out period represents a dangerous time for alcoholics, a
time when they are very susceptible to relapse. It is a time of great
physiological and psychological instability, a time when they are beset
by powerful urges, cravings, and yearnings, when their minds and
bodies alike scream out for alcohol as the cure for their distress or, if
not alcohol, then some related drugs, such as antianxiety medications
like Valium or Librium, barbiturates, or tranquilizers, which are commonly dispensed at hospital emergency rooms or alcohol detoxification centers. It is also a time when they remain relatively vulnerable
to the temptation to drink, not yet having established alternative habits
or protective patterns of thought against drinking.

Once they survive the ordeal of detoxification, their next obstacle to
recovery involves withstanding the generalized unease that is part of a
protracted alcohol withdrawal syndrome which can last much longer
than most ever imagined. Individuals can remain irritable, restless,
and depressed as an expression of abstinence from alcohol for up to
four to six months, and have trouble sleeping and difficulties in thinking for up to a year.[3] Because of their subtlety, these symptoms often
go undetected. Individuals are vaguely aware that they don't feel right,
but they don't know why. So they search for reasons. They blame
their work, their spouses, their circumstances, or even life in general
as a way of making sense of their feelings. Craving alcohol again is an
inevitable result of this way of thinking.

In light of these and other difficulties, it is remarkable that any alcoholics actually manage to recover. Yet a significant number do.[4] The most exhaustive literature review on this topic indicates that cumulative rates for "spontaneous recoveries" over a period of years—that is, recoveries that are not attributed to formal treatment or Alcoholics Anonymous but occur for other reasons, such as quitting drinking on one's own—vary widely from 10 to 42 percent in different reports, with yearly rates ranging from 1 to 33 percent.[5] The recovery rates after formal treatment appear to fare no better. The results of a typical eighteen-month follow-up study of treated alcoholics, for example, reveal that progressively smaller percentages are able to remain abstinent from alcohol over the entire time. About one-half of the alcoholics manage to stay dry for a minimum of three months; about one-third for six months; about one-sixth for twelve months; and less than one-tenth for the entire eighteen-month period.[6] The cumulative rates of relapse, which represent another way of looking at these same results, are even more revealing. After exposure to an intensive one-month hospital treatment program with abstinence as its goal, about 70 percent of alcoholics relapse during the first three months after discharge—the time of greatest physiological instability—and, after that, the rate steadily continues to climb to more than 90 percent by eighteen months.

What these discouraging, cumulative group statistics do not reveal is the typical pattern of drinking among these treated individuals. Once alcoholics slip or relapse, the course is not an inexorable, downhill one of steady and progressive drinking. Instead, it tends to be a roller coaster course of going "on and off the booze," during which alcoholics drink until they can drink no more, then remain dry until they can stay dry no longer. As a result, about one-half of alcoholics are dry at any given point in time.[7] This typical pattern is responsible for the artificially positive results reported in many uncontrolled treatment studies—the common 40 to 60 percent "recovery" rates—that do not take into account the fact that most of the alcoholics who are dry at any follow-up period will be indulging again shortly, while others who are drinking will be sobering up shortly, thereby keeping a relative constancy to these recovery rates. In their efforts to stay dry, most alcoholics manage quite well to stop drinking; they are just unable to keep from starting up again and perpetuating this cycle.[8]

This on-and-off drinking pattern has other features as well. Being dry with respect to alcohol does not mean that alcoholics are avoiding all other forms of drugs. During this difficult and highly vulnerable time, many become dependent on other chemicals—under a physician's care or on their own—substituting tranquilizers, stimulants, barbiturates, antidepressants, marijuana, cocaine, narcotics, or other psychotropic drugs for their beloved potion, trading one altered state of mind for another.[9] This may temporarily keep them from drinking, but sooner or later they revert. Unfortunately, as long as alcoholics tend to rely on mind-altering chemicals to suppress their craving—especially chemicals that mimic some of the effects of alcohol—the embers of craving remain alive, easily rekindled under certain circumstances, as when the substitute drug is stopped.

The experiences of Eddie H. Lee, an Air Force captain, is representative.[10] Two months had elapsed since his last drink. Captain Lee was giving a talk in church, to be followed by Communion. He passed the wine around and awaited his turn to drink. He drank the wine, then began feeling nervous and uneasy. When everyone left the church, he went into the back office and finished off the balance of the wine bottle. The old craving came back with a fury. He had to have more. He went to the officers' club and began ordering double bourbons. He became so drunk that he finally had to be taken to the hospital. Sometime later, after several weeks of abstinence, he caught a cold and went to the dispensary for cough medicine. He was given some liquid with codeine. He swallowed some and began feeling better. He had another swig, then another. He couldn't stop taking it. It had a narcotic effect on him, making him dizzy and woozy. When he finished the whole bottle, the terrible craving for alcohol came upon him again, similar to what he had felt when he took the Communion wine, and caused him to head downtown to the nearest bar.

Aside from a susceptibility to other chemical triggers, another feature of abstinence needs to be mentioned. Although it is usually the case that abstinence improves the relationships between alcoholics and their close friends and relations, this is not always the case. As bad as many alcoholics can be when drunk, they can become even worse when dry, demonstrating the need for attitude changes as part of the recovery process. This became apparent to me years ago when, as director of an experimental treatment program for alcoholics, most of the complaints I received from parents and spouses of alcoholics were not that they were still drinking, but that they had become impossible to live with now that they were dry. Some even went so far as to suggest that we should undo what we had wrought.

These observations are consistent with the results of a large-scale study of alcoholics who were abstinent for at least one year at the time of follow-up.[11] Of these abstinent alcoholics, about one-fifth, who had achieved more fulfilling lives and made reasonably mature adjustments to society, attributed their recoveries entirely to their own initiatives or to their affiliation with AA. But the remainder, though no longer drinking, suffered from all sorts of debilitating emotional problems, sometimes to a psychotic degree, or seemed to be inadequate personalities who led shrunken and constricted lives, avoiding all risk and novelty and functioning in very narrow ruts. In their case, abstinence per se bore no relationship to mental health.

What all these findings reveal about the long-term prognosis for alcoholism is discouraging. Only a small percentage of alcoholics manage to remain abstinent over the long haul, and of those who do, a comparably small percentage can be expected to achieve the more ambitious goal of sobriety. Why has the smorgasbord of conventional medical, psychiatric, or psychological approaches for alcoholism— whether they be individual insight-oriented therapies, cognitive-behavioral psychotherapies, group therapies, behavioral therapies like assertiveness or relaxation training, Antabuse, aversive treatments, psychotropic medications, or multimodal treatment programs—failed for the vast majority of alcoholics?[12] If that can be determined, then perhaps the necessary corrections can be made.

Throughout time, one of the standard treatments for addictive drinking has been to use some form of punishment, with the rationale that if the drinking became associated with an awful or unpleasant experience, the individual would desist from drinking in order to avoid this learned and conditioned consequence. One might suppose that these aversive therapies, as they are known, would be very effective, especially for highly motivated alcoholics, but this is not necessarily the case. These conditioned punishment techniques appear to have had as much deterrent effect on further drinking as the former use of public hangings on deterring pickpocketing, as evidenced by the famous Hogarth print in which pickpockets were circulating among the crowd of onlookers.

For a period of time, drugs like apomorphine or emetine were widely employed in many treatment centers to induce nausea, stomach cramps, and vomiting shortly after individuals smelled or tasted their favorite alcoholic beverage. Alcoholics underwent these experiences numerous times and sometimes received "booster" treatments at regular intervals after hospital discharge, with the expectation that this would prevent further drinking, because further drinking would be associated with being sick. While some treatment programs report impressive success rates with this method, and it continues to be used widely in parts of Europe and the Soviet bloc countries,[13] its use throughout the United States has declined.

For a period of time, painful electric shocks were applied to the forearms of alcoholics as they sniffed and tasted their favorite drink, but that approach, too, has essentially been discontinued.

Of all the aversive therapies, perhaps the most dramatic involved the use of succinylcholine, a curarelike drug. It is worthwhile describing this experimental approach since it not only exemplifies all aversive techniques but probably illustrates many of their basic deficiencies as well. For the program in question,[14] the physicians administered succinylcholine by intravenous drip to a group of alcoholics immediately after letting them handle, sniff, and taste their preferred alcoholic beverage. This induced a sudden respiratory paralysis in the alcoholics, along with a smothering feeling and a horrible sensation of dying. Not surprisingly, individuals exposed to this treatment not only recoiled when requested to smell and taste their favorite alcoholic beverage again, but displayed marked galvanic skin responses at those times, indicating that the conditioning program had been successful. If any aversive treatment should work, this one seemed to have all the prerequisites.

After reporting these impressive results, the investigators undertook a more extensive follow-up after hospital discharge[15] to determine how well the group classically conditioned to the drug fared in comparison with two other control groups—one in which the succinylcholine was given, but not in conjunction with the taste of the alcoholic beverage, and one in which only saline solution was given intravenously as the alcoholics placed the glass of alcohol to their lips. About two-thirds of both the alcoholics who had received the experimental treatment and those who had received the succinylcholine but not in conjunction with the alcohol continued to show distressing physical and mental symptoms, such as nausea, muscle pains, tremor, headache, anxiety and disgust, when approaching an alcoholic beverage. But this did not necessarily prevent them from drinking heavily again. They remained traumatically sensitized to alcohol, often repulsed by it, but not enough so that it significantly mattered. The explanations the alcoholics gave for their drinking were illuminating. When first experimenting with alcohol after the treatment, some attributed their unpleasant experiences to bad hangovers and continued to drink on. The very anxiety and fearfulness that others felt in the presence of alcohol seemed to serve as excuses for their drinking. Some found themselves aversively conditioned to the mixer more than the liquor itself. Others indicated that the induced aversion to their preferred alcoholic beverage simply caused them to switch to another kind of liquor. Yet others offered other reasons. Thus, what should have been a highly effective treatment essentially proved to be a failure.

What was wrong with these approaches to treatment? It is difficult to be certain, but a number of possibilities suggest themselves. The most obvious is the motivation of the alcoholics, but that does not seem to be the case. Individuals subjecting themselves to such drastic procedures as gastrointestinal distress, electric shocks, or respiratory paralysis must be reasonably motivated (or excessively masochistic), so some less obvious factors must play a role. One factor appears to be the highly artificial nature of the treatments and the unnatural settings in which they are administered, making generalization of the conditioned response to real life unlikely. Whatever is learned in a hospital, clinic, or laboratory setting tends to be relatively confined to that setting. It is also likely that the individuals were not conditioned against the usual times when they would most likely crave a drink—for example, at home alone or in a bar—and had not been taught any techniques for coping with any unexpected urges to drink. Punishing certain aspects of drinking behavior within a hospital, clinic, or laboratory setting not only may overlook other specific triggers for drinking but may also offer no protection against the actual craving for alcohol once it does arise.

Another problem with the aversive therapies is that they may not adequately promote the process of *extinction*—the gradual disappearance of a specific behavior like drinking—which is an important feature of any successful recovery. According to learning theory principles, any learned behavior should increase in frequency if it is rewarded (called positive reinforcement) and decrease in frequency if it is punished or not positively reinforced. Individuals stop performing any given behavior when there is no longer any incentive to continue. This extinction, however, will not likely take place when the given behavior is positively reinforced on an occasional, irregular basis (known as aperiodic reinforcement). In fact, this aperiodic reinforcement is likely to strengthen the original behavior, often far more than with consistent, positive reinforcement, since individuals tend to work harder to get the reward, never taking it for granted since they can never be certain when it will happen again. In a way, the craze for gambling may stem from this kind of reinforcement.

The problem of aperiodic reinforcement is central to the relapse process in alcoholics, making extinction of craving extremely difficult. As long as alcoholics occasionally continue to "slip" after a period of abstinence and, when they do, experience the desired effects (such as, intoxication or relief of withdrawal effects), the compulsion to drink is likely to remain as strong as ever. Since the alcoholic never quite knows when and if he will give in to one of his urges, his urge to drink will continue to plague him as long as the possibility of drinking again continues to exist.

For the extinction of craving to occur, alcoholics must somehow find a way to cope successfully with the urge to drink. Each time they successfully manage to avoid consuming alcohol when they desire it, they weaken its hold over them. In time, the continued avoidance of alcohol theoretically should result in weakening the compulsion for alcohol and in its eventual extinction, especially if another response is substituted for drinking. By definition, if the urge to drink is never satisfied, then it has no power over the individual.

These principles of learning theory have been applied in a deliberate fashion in the treatment of narcotic addiction,[16] but they, too, have yielded disappointing results. Narcotic antagonist drugs, such as cyclazocine and naltrexone, which eliminate the euphoric effects of narcotics by blocking opiate receptors within the brain, have been administered daily to drug addicts, who continue to have the option of "shooting up dope," as they always have, but now without the prospects of getting "high." Theoretically, the narcotic "hunger" should be extinguished in time as addicts discover that they no longer experience the same pleasurable effects from narcotics. This is an excellent illustration of nonreinforcement in action. The only problem is that most addicts refuse to take the narcotic antagonist drug on a voluntary, regular basis, and therefore, can never really experience extinction of their cravings.

Unfortunately, there are no pure alcohol antagonists comparable in action to narcotic antagonists that can make alcohol taste and seem like brackish water and thereby neutralize the compulsion to drink by the process of extinction. But even if such an antagonist drug were available, the problems would probably be the same as those for narcotic drugs. Alcoholics determined to drink would soon find an excuse for not taking it.

Perhaps the most common type of pharmacological protection available against alcohol involves the daily administration of disulfiram, also known as Antabuse, which can instill fear in individuals about the effects of drinking.[17] Antabuse acts mainly by blocking an important step in the metabolism of alcohol, resulting in a rapid, toxic buildup of acetaldehyde. As long as alcoholics have no alcohol in their systems, Antabuse has relatively innocuous effects. With the consumption of alcohol, however, individuals can expect to get violently ill and experience such symptoms as throbbing headaches, nausea, vomiting, palpitations, precipitous drops in blood pressure, and an intense fear of dying. The therapeutic advantage of Antabuse is that it stays in the bloodstream for about three to ten days after individuals have taken the last dose. So even if they stop taking the drug, they should not be able to drink right away—unless they are willing to risk feeling miserable. This prevents them from acting impulsively.

There can be no doubt about the therapeutic usefulness of Antabuse for selected, highly motivated alcoholics. For short-term management, the fear of drinking offers alcoholics that extra bit of protection and security at a time when they are emotionally shaky and physically on the mend. The problem, of course, with Antabuse, and probably the reason why it is not used even more widely, is that the alcoholic never really has the opportunity to gain self-confidence about his or her ability to resist alcohol without benefit of this pharmacological "crutch." Even more, while Antabuse makes it unpleasant to drink by "punishing" drinking behavior, it does not remove the urge. And the deterrent value of the drug is of little value if someone is determined to thwart it. This personal account by an author illustrates these matters well.[18]

"I began to take my Antabuse faithfully and was, through necessity, staying sober, but I was very unhappy. I was being forced to remain sober while the craving to drink was still upon me every waking moment. I couldn't eat, I walked the floor at night—couldn't sleep. My nerves were on edge. One morning when my wife gave me the Antabuse, I just pretended to take it, cupping it in my hand instead of swallowing it. I would have to do this for three mornings before I could take a drink. I wondered if I could wait. On the second day of skipping the Antabuse I decided maybe I wouldn't wait the required three days—maybe in my case two days would be sufficient.

"So I went into town and bought a pint. I figured just a pint wouldn't bother me. Going into a pool hall bar I asked for a chaser. I downed the first drink. Nothing happened. I tried another. Still nothing happened. Maybe the doctor was just telling me that story to scare me into not taking a drink. I finished the pint. Then it came upon me. My face began to flush. I could feel my pulse beating faster and faster. Red splotches broke out all over my body—my heart began to pound and pound, like it was laboring for the next beat. I knew I was going to faint. I told the bartender to get me a cab quickly, I needed to go to the doctor. He could see I did. The driver went as fast as he could with the traffic holding him up. My stomach began to ache, I was in terrible pain all over. I had to vomit, so I let the window down, and hanging my head out, I let it fly. Vomit flew back into my face, but I couldn't stop."

Though many relapsing alcoholics simply stop the drug and wait for the danger to pass, more impatient ones have learned how to inactivate the alcohol-Antabuse reaction by sipping small quantities of alcohol or, when desperate enough, simply drinking right on through the reaction even though they get miserably sick. The sicker they get, the more they continue to drink. Given this situation, I suspect that even if the experimental, sustained-release Antabuse pellets, which can be surgically implanted under the skin, were perfected for general use, alcoholics bent on returning to drink would continue to display similar perseverance and ingenuity in overcoming their deterrent effects.

What about the matter of psychodynamic, insight-oriented, or cognitive-behavior psychotherapies, conducted individually or in groups? On the assumption that excessive drinking represents a symptomatic expression of an underlying conflict or a reflection of irrational attitudes, these therapies attempt to get alcoholics to understand the genesis of their problems and, in the process, to develop more mature ways of responding to them.

Despite occasional successes, these approaches to therapy miss an essential point. The point is that, regardless of the alcoholic's original reasons for drinking, the drinking itself soon takes on a life of its own, relegating all other problems to minor roles. As the saying goes, first the man takes the drink, then the drink takes a drink, then the drink takes the man. Until the habituating or addictive aspects of alcoholism are addressed, all other treatment tends to be superfluous.

One of the problems in treating drug addiction or alcoholism with verbal therapies alone is that what an individual knows or learns under one condition or state of consciousness may not be readily transferable to another state of consciousness. This was illustrated well in the 1930s by Charlie Chaplin. In his movie, *City Lights*, Charlie, the tramp, encounters a drunken millionaire about to jump into the river and eventually convinces him not to. In gratitude, the millionaire takes the tramp back to the mansion and regales him with the finest comforts and refreshments before both fall asleep in the luxurious living room. In the morning, the millionaire, on awakening, does not recognize the tramp snoozing across from him on the sofa and instructs his manservant to throw him out. Charlie, naturally, is devastated and puzzled by this action, but in his pathetic and hopeless way, takes this rebuff as a natural happening in his daily life. Some weeks later, as Charlie is wandering down a street, the same millionaire, obviously intoxicated, comes bursting through a café doorway, surrounded by a crowd of friends. In his inebriated state, the millionaire recognizes Charlie, ebulliently embraces him as a long-lost buddy, and escorts him home again to the mansion, whereupon the same sequence of events recurs.

What this old silent movie so brilliantly manages to depict, of course, is what has now become known as "state dependent learning," a situation in which information learned in one state of consciousness may not be readily transferable to other states of consciousness.[19] This phenomenon may also explain why individuals easily forget dreams on awakening but may recall them sometime later during sleep, or even more to the point, why intoxicated alcoholics, who hide bottles of liquor in the closet one night cannot remember where they hid them the next morning, but can immediately remember after having a couple of drinks at a nearby bar.

In psychotherapy, the phenomenon of state dependent learning could account for why intoxicated alcoholics usually do not follow through on their avowed intentions to quit drinking, once they sober up, no matter how adamant their intentions, and why sober alcoholics may remember little about what they have learned from psychotherapy once they get some alcohol in their bloodstreams. At that moment, with the triggering of craving, all prior insights and learning about why they do what they do and all their good intentions about not doing it any more begin to vanish, and their new goal, which is the same as it always was when they drank, becomes like the character of Brick in Tennessee Williams' play, *Cat On A Hot Tin Roof*, who wanted to keep drinking until he felt the "click" in his brain.[20] If state dependent learning negates some of the beneficial effects of psychotherapy, two implications for treatment are clear. Either psychotherapy should be directed toward treating alcoholics not only when sober but also during different degrees of intoxication in order to ensure that what they presumably learn will be better retained in the event of a "slip," or else it should be directed toward keeping them from taking that first drink—a strategy adopted by AA—so that they needn't confront the problem of reneging on their good intentions.

Of course, the attractive but controversial hypothesis of state dependent learning is probably overly simplistic, ignoring such alternative explanations as the disinhibiting effects of alcohol on the frontal cortex of the brain (which could explain why conscience has wryly been defined as "a substance readily dissolvable in alcohol"), the influence of different discriminative stimuli as determinants of specific behaviors, and the ability of alcohol to unleash a whole array of conditioned responses. Whatever the true explanation, the problem remains clear. Virtually all psychotherapeutic approaches eventually falter once alcohol begins circulating in the brain, unlocking a highly predictable repertoire of the attitudes and behaviors associated with drinking.

Another deficiency with most conventional treatment programs, whether involving individual psychotherapy with a private counselor or therapist or intensive individual or group treatment within a hospital, is that no amount of talking can substitute for dealing with the problem as it arises in a real life setting. Just as state dependent learning tends to be limited to a given state of consciousness, information learned in one setting, particularly an artificial and unnatural one such as a hospital or shelter, may not be generalizable to situations encountered in real life. Whatever marvelous insights and intentions the alcoholic experiences in the world of empathetic nurses, caretakers, and physicians who try to take care of all his creature comforts and needs, may become useless or irrelevant outside the treatment setting once he encounters a complaining spouse, an unappreciative employer, or a favorite drinking buddy.

There is another potential problem with insight-oriented approaches. While insight into one's motivations may represent an effective method for dealing with a variety of neurotic problems or emotional conflicts, it seems to have little impact on the compulsion to drink.[21] This was pointed out to me most vividly in the late 1960s and early 1970s when LSD and other hallucinogenic agents, such as mescaline or psilocybin, were being touted as revolutionary, miracle cures for alcoholism. Caught up in the enthusiasm of the times, I, too, began experimentally administering these "psychedelic" (mind-expanding) drugs, as they were then called, for treatment. Sure enough, the results were miraculous. A substantial number of alcoholics exposed to various types of LSD treatments reported wondrous insights, integrative revelations, transcendental experiences, increased self-confidence, a sense of rebirth, and a strengthened motivation for sobriety, not just while they were under the influence of the drug but for many days, weeks, and months afterwards. There was only one catch. These remarkable insights bore no relationship to their other basic attitudes or how they lived later outside the hospital. Even more important, the individuals who received these treatments did not drink any less or function any better in society than those alcoholic individuals in a control group who, instead of the LSD, were only given a pad of paper and told to write about themselves for a two-hour period.[22] The impracticality of these drugged insights is depicted in a poem written by one of my alcoholic patients some time after receiving the experimental treatment.

Me and LSD

or

Lines of Junk by a Drunk

To me the bottle was Divine;
But many others came entwined.
Shiftless, sinful was my life,
Causing trouble, unknown strife;
Letting liquor take its grip.
But through God's will and LSD,
A want to live possesses me.
As of now I'm "on the wagon"
Joining world, not just draggin'.

Who Knows?

I woke one morning to the trill
Of a bluebird on my sill.
Good morning world! Oh, what the heck,
I slammed the window, broke his neck.

Another experience with a more conventional type of insight-oriented psychotherapy illustrates other potential limitations. Frustrated and demoralized by the numerous hospital readmissions of alcoholics who had been exposed to our intensive, multimodality, AA-oriented treatment program—many of whom I had been sure were "cured"—I decided in the late 1960s to investigate one of the key treatment components of our program: group psychotherapy. My intention was to study communication patterns among patients in order to determine what concerned them the most over the course of the treatment sessions. Six highly motivated alcoholics who had recently been detoxified at our alcohol treatment facility were selected for the group. The group members met regularly in a small room and seated themselves at a round table, about five feet in diameter, for 1½-hour sessions twice a week over the course of five weeks. As with ordinary group therapy, my approach was essentially nondirective and supportive, oriented toward promoting self-understanding and insight into the reasons for their behavior. To this end, individuals were expected to talk about whatever seemed important to them, without any specific instructions or constraints. The only new wrinkle in this relatively conventional format was that before each session I placed a full bottle of bourbon, a bottle of gin, a bottle of rum, a six-pack of beer, a container of ice, tongs, and six empty glasses right in the middle of the table around which we sat, fully prepared (and authorized) to let any or all of them drink if they decided to.

From the start, this was a particularly lively and talkative group. Individual members were exceptionally open about their failings and shortcomings and supportive and understanding about those in others. Important insights were experienced and excellent advice given throughout the sessions. One member confessed that he now saw how his "old lady" had driven him to drink, and the others commiserated. Another discussed his problems at work, and the others nodded in sympathy. Another talked about his terrible upbringing, and the others were empathetic. And all six mentioned the terrible things they had done to their families and how they had messed up their lives because of their drinking, and they all showed understanding. There was only one important topic lacking in all these meaningful discussions. Like the people Gulliver encountered on Laputa, who had one eye directed inward and one skyward so that they couldn't see the nose before their face, no member of the group mentioned the alcohol on the table, within easy reach and right before their eyes, during the entire fifteen total hours of group psychotherapy! [23]

All types of explanations for what happened are possible—the alcoholics were responding to implicit constraints, they thought they were being tested, it was all a big game—but this experience taught me an important lesson. Just because people meet in groups and talk about themselves in meaningful and emotional ways and afterwards claim benefit from this activity does not mean that they are really discussing what needs to be discussed and confronted. At least in Alcoholics Anonymous meetings, the compulsion to drink—and what to do about it—is of foremost concern and is brought up again and again. New members may choose to look about, offer excuses, deny their addictions, or blame others for their troubles, but unlike the people on Laputa, they are constantly confronted with the situation right before their eyes—they are alcoholics who lack the ability to control their drinking. And in all forms of insight-oriented or cognitive behavioral psychotherapy, unless this basic reality is addressed, as it usually is in AA meetings, all other talk is meaningless.

But even when alcoholic individuals confront their problem—the serious nature of their addiction—that is only the first step in their recovery. Alcoholics Anonymous, while highly effective, is far from universally successful. According to one enthusiastic claim, the success rate of AA is 75 percent: 50 percent on the first attempt and half of the remainder on a later try.[24] This is probably too optimistic an estimate when other reports, based on more specific criteria, list recovery rates ranging from about one-third to three-fourths of alcoholics who take the program seriously,[25] but it is still impressive.

My own follow-up study[26] on alcoholics tends to offer support for the general effectiveness of AA. Regular attendees demonstrated an overall better social adjustment and drank less than irregular attendees or nonattendees. One of the surprising findings of this study, however, involved the pattern of attendance at AA meetings. Though AA encourages total abstinence, most alcoholics attending the meetings, even those doing so on a regular basis, continued to drink periodically, then sober up, an on-and-off drinking pattern typical of alcoholics in general. These observations are confirmed by other studies. Griffith Edwards and colleagues[27] found that 57 percent of AA members "slipped" at least once and 18 percent slipped five or more times over the course of their membership. Though "slipping" is tacitly discouraged by AA, it may well serve an important function for the organization by reminding members of their continuing vulnerability.[28]

One of the major advantages and attractions of the AA philosophy is its simplicity and practicality.[29] It recasts and interprets all of the alcoholics' psychological conflicts, emotional difficulties, low self-image, and problematic life circumstances in terms of drinking or abstinence.

Anxiety, depression, and anger are regarded as avoidable consequences of the urge to drink rather than as a result of conflicts encountered in daily life. The world and human experience thereby become predictable and manageable. AA provides a number of routine rituals, codified behaviors, and fundamental prescriptions for dealing with alcohol and coping with adversity, all of which are embodied in its well-known Twelve Steps.[30]

For those individuals who had been struggling futilely on their own, AA offers its collective strength and wisdom, and a remarkable sense of fellowship, as well as living inspirational testimonials from those who managed to recover. The alcoholic soon comes to realize that he can't fool these people. This is a group that cares, that knows, that understands, a group made up of people like him. Through identification with the group and acceptance by it, the alcoholic also comes to absorb some of its power.

The conventional wisdom associated with AA is nowhere more evident than in its 90/90 rule. Members newly abstinent are encouraged to attend ninety meetings during the first ninety days. Not only does this expectation address the high vulnerability for relapse during this time, giving individuals the organization's support during this critical period, but it also gives them something to do every day, a way of distracting their thoughts from the constant temptations to drink.

For most individuals, the process of affiliation in AA appears to involve a sequence of phases.[31] After hitting bottom, they usually make contact with someone in AA who orients them to the program. Within this accepting atmosphere, they come to adopt the label of "alcoholic" and then make a commitment to get seriously involved in the program. One of the most important parts of this commitment is the willingness to be judged by group standards and to accept subordination to the group. At meetings, individuals are expected to tell their stories and to listen to those of others about how bad it was when they were drinking ("drunkalogues") and how good it has become since they stopped ("sobriety stories"). The sense of group identity becomes strengthened even more when individuals begin doing twelfth-step work—that is, spreading the message to other alcoholics in need.

This personal account by a recovering alcoholic offers a description of what typically can take place in AA. "I went to a lot of AA meetings at first," he told me, "and I got a lot of reinforcement, a lot of stroking. I became part of that society. The others became my peers. I was accepted for the first time in my life. I gained stature and status in the group . . . by not drinking. Then I started feeling better about myself. It's like behavior modification. Going to AA meetings over and over again. Seeing what happens to other people. Seeing a guy with 17 years of sobriety fall off the wagon . . . seeing how sick he is. So how did I train my mind? How did my attitude change? It was like getting programmed, by going to the meetings. It's like a constant recharging of my emotional batteries. Every time I'd go to a meeting, it took less and less effort to stay sober."

Putting the issue of AA effectiveness aside,[32] we are left to wonder why more alcoholics do not avail themselves of its services, since its success rate is probably at least as good as that of any other approach and joining it is free.[33] Estimates are that only 5 to 10 percent of alcoholics in this country use AA, and as indicated before, of all those who recover, only about 10 percent do so through AA. The answer seems to be that not everyone is ideally suited to become a member of this organization and make the kind of commitment necessary to obtain optimal results. Because of its special nature, the organization seems best suited to highly motivated individuals with strong affiliative needs, who are willing to accept the principles and precepts embodied in the Twelve Steps and who do not rebel against the highly organized structure and expectations of the program. Individuals who rebel against an alcohol-centered existence for the remainder of their lives, who cling to the belief that they can change themselves, who believe that the open confession of their alcoholism to others may harm them professionally and socially, as well as those who react negatively to the spiritual orientation of the program, tend to avoid any involvement with AA. For example, such writers as Eugene O'Neill, John Steinbeck, Scott Fitzgerald, and Malcolm Lowry, all of whom had serious drinking problems, have shied away from this kind of affiliation.[34]

Though AA is unique in its exclusive focus on alcoholism, other quasireligious or lay organizations also appear to have success with this problem. Dramatic drops in drug and alcohol use, for example, have been reported for individuals joining the Divine Light Mission, the Unification Church, and other Jesus movement groups.[35] In one survey, 91 percent used alcohol and 97 percent used drugs prior to their conversion; afterwards, none used either. Organizations like Friends of Sobriety or Women for Sobriety, which are modeled after AA, also seem to help many. This suggests that the very process of group affiliation, particularly with one that maintains a strong prohibition against alcohol and drug use and keeps members busy with various activities, represents a powerful aid in the maintenance of abstinence.[36] Almost any "identity transformation organization" that shields individuals from outside physical, social, and ideological forces and tries to alter their attitudes and behaviors seems to have the potential to promote recovery.[37]

But these represent global observations about established treatments or helping organizations. They do not address the most important ingredient for successful recovery—the desire for help. Without a firm decision to quit drinking and remain sober, individuals are destined to perpetuate the cycle of drinking and then drying out, a never-ending roller-coaster ride, regardless of whether they receive treatment from others or try to make it on their own, until death or some equally dire circumstance intervenes. With this the case, the question then is: What does it take for them to develop the proper frame of mind for sobriety?

6

The Proper Frame
of Mind

In his remarkable autobiography, Jack London writes eloquently about the spell cast by John Barleycorn, a Mephistophelian personification of alcohol, who promises "maggots of fancy, dreams of power, forgetfulness, anything and everything. . . . He can tuck in his arm the arm of any man in any mood. He can throw the net of his lure over all men. He exchanges new lamps for old, the spangles of illusion for the drabs of reality, and in the end cheats all those who traffic with him."[1] What chance do alcoholics have against so formidable a foe?

For most alcoholics, it is far simpler to sacrifice body and soul for their consuming obsession than to give up the golden glow of intoxication, the sweet oblivion of drunkenness, and an entire lifestyle centered around drinking. Even as their world crumbles about them, even though their lives are left in shambles, alcohol still offers them the illusion of freedom and truth or, as William James, the psychologist and philosopher, remarks, the opportunity to make life seem *more utterly utter.* "Sobriety diminishes, says no; drunkenness expands, says yes."[2] Carl G. Jung, the psychiatrist, describes intoxication as the "equivalent of union with God."[3] No wonder the bondage to alcohol, a drug that allows alcoholics to be anything they want. The wonder is that any manage to break free.

Yet, as already noted, many do. Not necessarily when they should—before health is impaired, promising careers ruined, close relationships destroyed, and great personal damage wrought—but when they are ready and in the proper frame of mind. The crucial issue for rehabilitation is how to get the alcoholics to arrive at this point.

When abstinent alcoholics are asked why they stopped drinking, most credit special life events, such as getting sick, a divorce, or being fired—not their past treatment or hospitalization—as the reason.[4] But crediting life events with motivating power seems like attributing life-like qualities to inanimate objects. The difficulty is that the events themselves possess no intrinsic or fixed meaning. They only take on meaning through the perceptions and interpretations of individuals.[5] A given event can mean different things to different individuals or different things to the same individual at different times. The threat of divorce may come as a relief for one person and be utterly devastating for another. An alcoholic may shrug off the first bout of liver failure but become terrified by the next. Or he may have one car wreck after another, alienate his loved ones, experience bankruptcy, and become seriously ill as a result of alcohol, but only decide to quit drinking in response to a seemingly insignificant event, like betting someone that he could. All this suggests that it is not what happens at a particular time but how an individual reacts to what happens that is most influential in decisions to abstain. Take the following examples.

One night Marty became furious with his wife when she began badgering him about his drinking. He started to chase her around the room, but his sixteen-year-old son, who was wearing a neck brace because of a compressed vertebra from a football accident, tried to restrain him. He flipped his son over his back and made him land on the floor, almost rebreaking his neck. He was so shocked by what he had done that shortly thereafter he quit alcohol for good.

Dr. Johnson, a forty-two-year-old surgeon, continued to drink even more heavily after his divorce, the loss of a custody fight for his children, a bout of d.t.'s, and constant arguments with his partners about his impaired judgment and embarrassing behavior. Nothing could make him stop. Only when he lost his admitting privileges to the hospital and was threatened with revocation of his medical license did he become sufficiently motivated to quit drinking and turn to AA for help.

For Gunther, a sixty-year-old engineer, the situation kept getting worse. He was having troubles with his family, difficulties at work, and he felt physically terrible and disgusted with himself. Because of his severe diabetes and kidney trouble, he was warned by his physician that he would die if he continued to drink. Ignoring this threat, he got out of bed in the middle of one night to get some liquor when he saw his wife crying. He felt so guilty and awful over what he was doing to her that he decided to quit drinking.

Each of these men characterized his experience in similar terms— as hitting a personal low or "bottom"—even though the circumstances surrounding these decisive moments differed appreciably. Marty was not bothered in the least by his abuse of his wife but was shaken to the core by what he had almost done to his son; Dr. Johnson drank even more heavily after his recent divorce but was shocked into sobriety by the loss of hospital privileges and the looming humiliation of having his medical license revoked; and Gunther showed little concern over the sorry state of his health but was stunned by the sight of his wife crying. Not only does each person respond to personal events in different ways, but what moves one person toward sobriety may not affect another at all.

The experiences of these three individuals are not so different from those of other recovered alcoholics who first have to experience a low point or bottom—a personally defined crisis that they interpret as intolerable[6]—before they are ready to commit themselves to sobriety. These crises are usually of a physical, emotional, social, or spiritual nature. Individuals may experience life-threatening, painful, or debilitating illnesses, such as cirrhosis of the liver, pancreatitis, bleeding ulcers, heart failure, or delirium tremens, as a consequence of drinking, or just be "sick and tired of being sick and tired." Or they may encounter personal humiliation, public disgrace, failure, despair, or distress in response to divorce, job loss, rejection by family and friends, lawsuits, bankruptcy, accidents, or jail. As with the picture of Dorian Gray, they perceive themselves as uglier and uglier until they no longer can stand what they see.

Again, it is important to emphasize that these personal lows are not absolute thresholds or objective events for given individuals. They are highly subjective interpretations about what is unacceptable or intolerable at a particular moment in time. But hitting bottom does not mean things could not have been worse. Like archaeological digs that unearth one city beneath another, there is almost no limit to how bad the situation could potentially get as long as individuals are still alive. What may be tolerable at one time may be intolerable at another. It is up to individuals to define their own personal limits or thresholds— what it takes to get them to stop. The man who almost hurt his son could just as easily have held off quitting until some later time when he had actually broken his son's neck. The physician who lost his hospital privileges could have postponed his decision until he killed a patient while in a drunken haze. And the man who was moved by his wife's tears could have ignored them once again, just as he had done so often in the past, continuing his self-destructive course until his physical health deteriorated further or she took her life. Fortunately, each of these decided not to push his luck further, experiencing his crisis of conscience before even greater damage occurred.

The question is, what brings alcoholics to this critical juncture in their lives, especially after failing to remain sober so many times in the past or else deluding themselves that they had no real problem. Why do they opt for sobriety at a particular moment in time, not a week earlier or a week later, but right then? Where do they find this courage to change, the conviction that they can, or the knowledge that they must?

From careful questioning of recovered alcoholics, one answer seems clear. Most decisions to quit drinking are not the result of haphazard or unknown forces but of discrete circumstances or events which have special, personal significance for individuals at a particular moment in time. The timing must be right. Though sometimes seeming sudden and impulsive in nature, these decisions can only be understood in a historical context, representing a logical culmination of the person's past struggles with alcohol and his strivings for sobriety. There appears to be a subconscious ripening of motives, as William James writes, until some critical threshold is reached or a significant event occurs which prompts the pivotal decision. The decision has been building over time, ready to be implemented at some critical moment—in retrospect a turning point, a time of personal crisis and change, a time when the alcoholic chooses to do what he knows he must do.

Because individuals experience a low point or bottom does not mean that they will not experience others or that recovery will be assured. It is more the rule than the exception for individuals to experience a number of drinking-related crises over the course of their drinking careers, each crisis serving to launch a new period of abstinence of variable duration. Because of the tendency of recovering alcoholics to attribute their worst time to the start of their most recent episode of abstinence ("This time I *really* hit bottom"), comparatively discounting the personal significance or importance of similar times in the past, it is often difficult to identify the critical ingredients for the successful launching of sobriety. When the process of recollection tends to elevate prior low points to new levels, it is impossible to tell whether the most current bottom will later suffer the same fate.

This brings up another feature about bottoms. It is my impression, from the study of numerous drinking accounts, that many alcoholics are not aware that they are experiencing a personal bottom at the time it is happening. It may be perceived as just another in a series of crises or miserable times when it occurs, but they later label or interpret the situation as such when they are recovering in order to justify, in retrospect, their decision to quit drinking. All new behavior needs a beginning, a dramatic moment, to indicate the transition between what was before and what came afterwards. The designation of "bottom" serves as a convenient, retrospective label for a turning point.

It would be wrong to imply that *all* alcoholics need to experience a low point or bottom before they can successfully change.[7] Sometimes alcoholics quit drinking not on their own initiative but because others pressure or force them to. An employer, for example, may insist that an employee receive treatment for his excessive drinking. A wife threatens divorce unless her husband seeks help. A physician pressures a patient to enter the hospital or go to Alcoholics Anonymous. An individual develops an allergy to alcohol, becoming deathly sick whenever he drinks it.[8] A judge issues a court order for hospitalization or sends the individual to jail. A doctor prescribes medicines that are incompatible with alcohol for an unrelated ailment. Or an individual joins the Navy and then ships out to sea.

The two cases below illustrate how quitting drink may sometimes not start out as a matter of choice.

At first Genevieve, a nursing anesthetist, drank only when she got home from the hospital, but soon she began bringing vodka to sustain her when she was on twenty-four-hour call. One day she convulsed at work, and the neurologist, after an extensive evaluation, suggested that the seizure was due to her drinking. She dismissed the notion and began drinking even more heavily. Weeks later, a perceptive physician at the hospital noticed her intoxication and ordered her to leave. He told her that she wasn't in any condition to work and that she shouldn't come back until she did something about herself. She expressed indignation about this to a girlfriend who, fortunately, urged her to seek help. She made an appointment to see a psychiatrist later that week, but continued to drink steadily up until that time. During the initial session, she tried to convince the psychiatrist that her problem was entirely emotional, that she drank because of stress, but he insisted that she was an alcoholic and said he would not see her again unless she joined Alcoholics Anonymous. So she did call AA, not because she believed she was an alcoholic, but, as she later confessed, to con her bosses. If she could say that she went to a psychiatrist and to AA, they would have to let her continue working. Before attending her first meeting, she decided that she would stop drinking for a time to prove to everybody that she wasn't an alcoholic. Then she could return to drinking. But after she went to several Alcoholics Anonymous meetings, she realized that she had a serious problem with alcohol. And the only way for her to deal with it was to try to quit drinking permanently.

Then there is Shecky Greene, the comedian, who would go on binges every three to six months. When he drank, he lost complete control, and often was publicly humiliated, or even arrested. Afterward, he would become depressed, often suicidal. This pattern continued until he became seriously ill and required surgery on his parathyroid gland. After the operation, the doctor prescribed a medicine that Shecky would have to take as long as he lived if he wanted to get better. The medicine, though, was incompatible with alcohol. Because he wanted to live more than he wanted to drink, he opted for abstinence.[9]

Not all instances of abstinence from alcohol begin when individuals are emotionally, spiritually, or physically bankrupt or quit because they must. A sizeable minority, perhaps 25 percent, do manage to break the bondage of their addictions long before they have fallen from social or physical grace.[10] Of those who now join Alcoholics Anonymous, at least 50 percent supposedly do so without waiting to hit bottom.[11] Somehow they manage to take a hard look at themselves, see clearly where they are headed, perceive "the bottom of the well," realize that they are losing control over their lives, and resolve to change their destructive behavior before it has a chance to destroy them. These are people who have the capacity to raise their own personal low points—to lower their emotional thresholds to what is personally intolerable to their consciences—long before any serious damage has been wrought. In AA terms, these individuals have had the bottom come up and hit them.[12] An assortment of mundane events, such as a marital squabble, a new business venture, a spouse quitting drinking, or even the need to diet, rather than a serious illness or crisis, may contribute to the decision.

Because "raising the bottoms" of alcoholics, getting them to quit drinking before they suffer irreparable harm, represents a goal of all those concerned, several examples of how this can come about should prove enlightening.

Gary, a fifty-year-old law enforcement officer, was sipping a cocktail and reading a book when he was approached by his son, who asked without recrimination, "Are you an alcoholic?" He pondered the question for a long while, wanting to be completely honest, then simply answered, "Yes." His son said, "Why don't you get some help?" Gary dearly loved his son. He decided then and there to do something about his drinking. He thanked his son for his concern, then made arrangements to enter a hospital.

Joshua, a professor of music at a university, began a five-year period of sobriety after making a New Year's Eve bet that he could quit alcohol longer than his friend.

It took a rather startling experience to cause Duane, a claims representative for an insurance company, to wonder "Now what in the hell are you doing here?" One night, at a bar he frequented, he witnessed two men fighting about one of the dancers. One man pulled out a knife and slashed the other across his belly. The injured man was bleeding profusely, his viscera hanging out from the wound. It was a gory spectacle. Though he had consumed five beers, Duane instantly sobered up. While everybody in the crowded room stood around, he tried to help. He got some ice, stuck it on the man's belly, and tried to keep the intestines from oozing out all over the floor. As he waited for the ambulance to arrive, he kept asking himself over and over just why he patronized joints like this and why he needed to drink.

Dr. Prescott, an eminent surgeon, had just gotten back from the liquor store with a pint of vodka, took four quick swallows, and thought, "It will be just about fifteen minutes before these damned shakes are gone." At that moment, insight struck, and he realized that he had become physically dependent upon alcohol. His self-esteem was severely shaken by this shocking discovery, the awareness that he was lacking in control. That was when he decided to quit drinking.

It is important to emphasize that these cases, while of theoretical and practical interest, should not detract from the fact that the surest embarkation point toward sobriety for the vast majority of alcoholics is still a personally defined low, or bottom. But to recognize this does not explain why alcoholics need to be so miserable, to perceive themselves at the bottom of an abyss, before they can become motivated to change their self-defeating behavior. Why is it that reasonably intelligent men and women remain relatively immune to reason and good advice and only choose to quit drinking when they absolutely must, after so much damage has been wrought? What is there about alcoholism, unlike any other "disease" in medicine except certain drug addictions, that makes being in extremis represent a potentially favorable sign for cure?

Undoubtedly, the potent pharmacological effects of alcohol itself must play an important role in perpetuating continued drinking. Almost up until the bitter end, even when the euphoric effects of alcohol become more fleeting and short-lived, it can still be relied upon to help the alcoholic feel more "normal." Like the alcoholic proconsul in Malcolm Lowry's *Under the Volcano*, the drinker tries to strike "the fine balance between the shakes of too little and the abyss of too much." But beyond the habituating or addictive aspects of the drug, certain psychological features of the alcoholics themselves also need to be taken into account. It is these very features that help explain why alcoholics usually do not heed the warnings and entreaties of others, why they usually resist the help offered even after they asked for it.

At the outset, one fact needs to be made very clear. Despite the countless claims about certain characteristic personality patterns that predispose individuals to alcoholism—dependent, impulsive, immature, and hostile traits are usually incriminated—available evidence argues to the contrary. There is no such thing as a typical alcoholic personality. After a review of the voluminous literature on this topic,

Mark Keller, former editor of the *Journal of Alcohol Studies*, concludes wryly that the investigation of any personality trait in alcoholics will show that some studies say they have more of it, and some say less of it than nonalcoholics, or, rephrased, that alcoholics are different in so many ways from nonalcoholics that it makes no difference.[13] There are also probably as many different innate personality patterns among alcoholics as can be found in the general population.

But the lack of a typical alcoholic personality does not mean that there is not a constellation of inchoate attributes common to alcoholics, or most individuals for that matter, which can become exaggerated and hypertrophied in response to a growing dependence upon alcohol and all of the deception, interpersonal difficulties, and expenditure of energy it entails to maintain so destructive a habit.[14] Dr. Harry M. Tiebout, an early authority on alcoholism, made certain observations in the 1950s about the defensive armor of alcoholics, which, though overly simplistic, seem just as pertinent now as then. According to Tiebout, the operation of two automatic, unconscious defense mechanisms, defiance and grandiosity, accounts for the unwillingness of alcoholics to secure and accept help.[15] Defiance allows the alcoholic to manage his anxiety by ignoring or denying the reality of his growing lack of control. At times this denial can reach almost psychotic proportions, especially when it becomes global and immune to reason. Even when confronted with incontrovertible evidence about his excessive and harmful drinking, the alcoholic usually prefers to remain blind to this reality than risk the prospect of not being able to drink. The companion response of grandiosity, a key component of "alcoholic pride," only reinforces this attitude by allowing the alcoholic to maintain the fiction that he possesses the ability to regulate his drinking and that his life essentially remains under control. As his situation deteriorates, the alcoholic frantically mobilizes a whole array of other psychological defenses[16]—rationalization, distortion of reality, projection—that allow him to excuse his behavior, bend the truth to suit his needs, and blame others for his predicament in order to perpetuate the illusion that he is perfectly alright and can continue to drink. The essential purpose of all these mental maneuvers is to allow the alcoholic to do what he wants without suffering personal guilt.

These are formidable defenses, impenetrable and unshakable to most assaults by reality. No wonder that cataclysmic psychological events, physical shock waves, or volcanic emotional upheavals often become necessary to shatter the alcoholic's complacency and reshape the landscape of his habitual attitudes and values so that he can begin to entertain the heretofore unimaginable notion that he not only must stop drinking, for he probably has managed to do that for days, weeks, or months in the past, but that he must never *start* drinking again.[17] That is the critical test.

For Alcoholics Anonymous, the personal lows represent welcome events, propitious states of mind,[18] for it is then that alcoholics become more willing to take the first step toward sobriety by admitting that they are powerless over alcohol and, as a result, that their lives have become unmanageable.[19] In a remarkably perceptive statement to the New York State Medical Society, Bill Wilson, cofounder of AA, revealed that once

> the individual has accepted the fact that he *is* an alcoholic and the further fact that he is powerless to recover unaided, the battle is half won . . . "he is hooked." He is caught as if in a psychological vise. If the jaws of it do not grip him tightly enough at first, more drinking will almost invariably turn up the screw to the point where he will cry—enough. Then, as we say, he is "softened up." This reduced him to a state of *complete dependence* on whatever or whoever can stop his drinking. He is in exactly the same mental fix as the cancer patient who becomes dependent, abjectly dependent if you will, on what . . . men of science do for cancer. Better still he becomes "sweetly reasonable," truly open-minded, as only the dying can be.[20]

Here are several examples.

Despondent about his divorce and the loss of his job, Tom finished the whiskey, then put the shotgun against his belly and pulled the trigger. Only after he entered the hospital did he experience remorse for what he had done and the anguish he had caused his family. On the operating table, he promised the surgeon that he wouldn't take another drop of liquor if he survived the operation. He managed to live, but he was left seriously disabled and in constant pain. True to his word, though, he never drank again.

Mary, a nineteen-year-old guitarist, had been having a hard time finding work. For a long while, she had been in and out of detoxification centers after episodes of smoking pot, taking hallucinogens, drinking heavily, and feeling sorry for herself. One day she was hitchhiking home to visit a friend and stopped off at a bar to use the bathroom and get a soft drink. She had already consumed almost two quarts of vodka and started nipping on a pint she was carrying with her. Suddenly she began having chest pains and blacking out. Things became hazy and she couldn't see well. "The next thing I knew," she said, "I was on the ground and people were shaking me and my chest just felt like it was going to blow up. They called an ambulance and took me to the hospital. They said I'd almost had a coronary. My heart had stopped actually, but it started up again itself." As soon as she sobered up, she made up her mind to quit "or I'm dead." That has been her primary motivation to remain sober ever since.

Roland, a safety director for a large construction firm, had just pulled out to pass a truck when he noticed a car coming toward him. With his reflexes slowed by alcohol, he veered too sharply, got his tires caught in the shoulder of the road, lost control of the car, and crashed head-on into another car. When he regained consciousness several hours later, he learned that one of the two women in the other car had to be rushed to the city for emergency surgery, while the other was in serious condition in the same local hospital he was in with smashed teeth, cracked ribs, a collapsed lung, and a broken leg. A compulsion to make amends overcame him. He dragged his bruised body out of the bed and hobbled down the hospital corridor until he found her room. Her husband sat near her bed. He introduced himself to them both, admitting responsibility for the accident, then asked if he could help in any way. They were nice enough, under the circumstances, but they did not grant him the absolution he was seeking. As he left the room, he realized that he was through drinking.

Janet, a fifty-five-year-old freelance writer, joined Alcoholics Anonymous only after she felt totally defeated, mentally and spiritually.

One incident after another began happening so that she could no longer deny the self-destructiveness of her drinking. Within a relatively brief span of time, she had totally wrecked three cars, was charged with a driving-while-intoxicated violation, and her husband threatened divorce. But it was the near drowning that proved decisive. "I got so drunk," she said, "that I decided I could swim across this huge, cold, icy lake. . . . I got in the middle and went down twice, and my life flashed before me. And I said, 'God, please don't let me die, this isn't how I want to die.' Ironically, as I went down, I found a rock and got out of the water, but I went into shock. That was the turning point."

Not all personal low points need be so intense and dramatic. Nor are the feelings of pain, fear, guilt, shame, grief, despair, humiliation, or powerlessness always defining characteristics. None of these feelings has been sufficient in the past to serve as a motivator for change. What is characteristic of the bottom experience of alcoholics is a special kind of anguish, one of realizing that they have sunk below certain basic, minimal personal standards of decency and behavior, and that they cannot live with what they have become. With the type of anguish, personal pride becomes bankrupt, rationalizations and denial fail, and alcohol no longer offers the blessed oblivion—it only makes the experience worse.

What is so fascinating about this propitious state of mind is its "sobering" nature. The alcoholic becomes keenly aware of the reality of his predicament. He no longer struggles to maintain the illusion of self-control. He is disgusted with himself. He has lost the fight. He feels defenseless, vulnerable, and utterly helpless, desperate for guidance and relief and ready to entrust himself entirely to an appropriate therapeutic entity, whether it be an alcoholism counselor, a treatment center, a shelter, a religious organization, Alcoholics Anonymous, or a personal God. With this act of "surrender,"[21] as with religious or ideological conversions, huge emotional displacements and rearrangements seem to take place. Ideas, emotions, and attitudes that were once guiding forces are suddenly cast to one side, and a completely new set of motives may begin to dominate.[22] It is as though the alco-

holic undergoes a "critical perceptual shift,"[23] suddenly seeing himself
and the world about him in an entirely new light, much like what
happens with the rapid perceptual changes in a Müller-Lyer illusion
in which an ugly old lady, with a blink of the eye, become an attractive young woman. Interested family, therapists, and helping agencies
that the alcoholic previously regarded as adversaries trying to get him
to quit drinking now become viewed as allies helping him to remain
sober.[24] Former drinking buddies and favorite haunts now become
people and places to be avoided. And everything about his former life
that once seemed glamorous and attractive is now regarded as aversive
and undesirable.[25,26]

It should not prove surprising that while in this frame of mind, this
state of utter despair and helplessness, many recovering alcoholics—
somewhere between 5 to 15 percent of those interviewed—should be
predisposed to religious or spiritual experiences.

"Man's extremity is God's opportunity," so the saying goes. For many
alcoholics, this saying is especially apt. Utterly defeated, they turn to
God as a last resort, as their only hope for sanity and sobriety.

In an oft-quoted letter that helped lay the foundation for much of
AA philosophy, Dr. Carl G. Jung eloquently stated why the belief in
God is so important in the struggle against alcoholism. "I am strongly
convinced," he wrote, "that the evil principle prevailing in the world
leads the unrecognized spiritual need into perdition, if it is not
counteracted either by real religious insight or by the protective wall
of human community. An ordinary man, not protected by an action
from above and isolated in society, cannot resist the power of evil,
which is called very aptly the Devil. . . . You see, 'alcohol' in Latin
is 'spiritus' and . . . the same word for the highest religious experience as well as for the most depraving poison. The helpful formula is:
spiritus contra spiritum."[27]

For Bill Wilson, the recipient of the letter, these words were almost
of revelatory import, for they confirmed all that he had experienced:
his unsuccessful attempts in the past to quit drinking, his resistance to
the notion of a greater Power involved in the affairs of humankind,
his desperation for help, and finally, the transcendent, spiritual expe-

rience that made his sobriety possible and contributed to his zeal to share his newfound insights with other alcoholics. *Spiritus contra spiritum.* These were particularly significant words, since he had come to believe that without the intervention of a personal deity, he and countless others like him in AA would never have been able to resist the evil force inherent in alcohol.[28]

The claim has been made that 10 percent of all AA recoveries happen through a "rapid awakening" by either a true religious experience or a sweeping psychological event. The other 90 percent attain the same result more gradually.[29] Even though those with gradual awakenings have not experienced the awesome presence of a divine being or a dramatic revelation, they nevertheless become increasingly convinced that their recoveries are dependent upon the protective interest of a Higher Power. These different types of recovery, interestingly enough, correspond well to the two major types of religious conversion: the *self-surrender* type, in which the personal will is given up in an abrupt, often dramatic fashion, and the *volitional* type, in which the regenerative change is gradual and consists of building up, piece by piece, a new set of moral and spiritual habits.[30]

Even in the absence of reliable statistics, the AA claim about the percentages of personal awakenings cannot be easily dismissed. Regardless of precise figures, it poses a number of important problems for anyone seeking a scientific understanding of alcoholism. Why should alcoholism, unlike any other "disease," be regarded as relatively immune to medical or psychiatric intervention and require, as AA principles insist, a personal relationship with a Higher Power as an essential element for recovery? Forget for the moment that the large majority of recovered alcoholics have managed to break their addictive habits on their own initiative or through more natural means, without invoking the intervention of supernatural forces. The point is that as long as logical explanations or scientific theories fail to account for these extraordinary, spiritual experiences, assuming these reports of subsequent recovery to be true, then alcoholism may be more properly regarded as a "disease of the soul" than as a biological, behavioral, or social disorder, presumed to be caused and eventually cured by natural means. The question, simply put, is how to make scientific or clinical sense out of these claims without impugning the integrity of the individuals involved or regarding them as misguided or deluded.

Here are some vignettes that illustrate the basic dilemma. Though the reported experiences all involve "rapid awakenings," they may be assumed to represent models for what happens with the far more common, gradual changes that presumably occur in slowly evolving spiritual convictions that end up at a similar point but which do not have the dramatic imprimatur of an initial divine intervention.

Ellen, a thirty-five-year-old writer, tried to assuage the guilt she experienced over her affair by drinking even more heavily. If she continued to stay drunk enough, she would not have to think about the pain she was causing her husband and children. One night, seated in the bathtub, she wanted to die more than anything in the world. Then "I heard a voice, and the voice said simply, 'Before you die take one realistic look at what your death will mean.' " All the times she had tried suicide before, she had rationalized that everyone would be better off without her—her children, her husband, everyone. But sitting there that night, she saw what it would really mean to her children if she died. "And I saw that they needed me, not as the *me* my family sees or the *me* as I would be, just *me* as I was, sitting there drunk and suicidal. That the love I had to give them was uniquely mine, and uniquely for them. . . . And that their lives would not be improved with my death. It would serve them terribly. . . . And then I said, 'I don't want to die, I want to live.' . . . I had never made a commitment to life, and I made one that night. . . . And I heard this voice again, and it said very simply, 'You can't stay alive and drink.' And that was how it happened." She called her husband and asked him to take her to the hospital, ending nineteen years of heavy drinking.

Barry, a thirty-seven-year-old contractor, felt that he was going to die, but he didn't care since he wouldn't have to commit suicide. He had bleeding ulcers, was on the verge of pancreatitis, and had to walk with a cane because of weakness in his legs. The day before he had been drinking heavily, so that when he got up that particular morning his first thought was to find some liquor or drugs. The day passed but he never got anything to drink except a couple of beers. That night he was afraid to stay by himself so he asked his wife to be with him. He was scared of reexperiencing the hallucinatory voices that tormented and harassed him. Then, as his wife fell asleep, he heard a voice which said, "Barry, this is your last time." "I thought, 'Now they are talking to me,' and I really got scared. Then the voice said, 'Barry, this is your last time.' I thought, 'I still don't know what they are talking about.' Then I heard this voice say, 'Ask my forgiveness.' I thought, 'Who?' . . . Then I heard myself saying, 'God, you've got to help,' and it sort of startled me because I remember saying that out loud. By this time I was worrying, I knew that my wife was going to leave me and I was going to die. . . . I told my wife, 'There's just got to be a way we can work this out' . . . even though she heard that many times before." So they went away for two days and that was the best time of his life. Since that night the desire for alcohol was gone.

Margie awoke that morning detesting herself, feeling as miserable as could be. Because of the terrible hangover and shakes, she went to the liquor store, bought several pint bottles of whiskey, and then headed toward the beach. She threw up in the public bathroom, then lay on the beach and began watching the waves. In a short while she went to the bathroom again and had the dry heaves. After a while she put her hand on the wall and said, "God help me," and began saying it over and over. Then she had a spiritual experience, along with a profound sense of commitment. She was going to ask for help again, but then realized that she already had been helped. She drove home in her car, looked up the AA phone number, and for the first time called for help. From that moment forward she remained completely sober.

Don Musgraves,[31] a writer, started reading the New Testament again, searching for the answer, then his eyes fell on I John 1:12. "I swallowcd hard and re-read those brief words: the print leaped up to me, suddenly communicating in a way other than words.

"Then, an explosion in my brain. A sudden white light *flooding* the darkness. I realized God was actually speaking to me, heard His voice saying, 'Receive my Son.'

"I was jarred so hard I shook. . . .

"I was hardly aware of time passing, as I read a few lines of the New Testament and repeated the prayer periodically, tears in my eyes, tears because I knew what kind of man I was and I was genuinely sorry.

"I kept crying and saying the *Sinner's Prayer.* . . .

"All through the darkness of that night, I tossed and turned.

"Daylight came. . . . Same old jail cell. Same old everything. Everything except me . . . me!"

Dr. Bill Jackson[32] was rapidly deteriorating into hopeless alcoholism. He had lost his wife, family, his job, and was near death physically. Then one day at the hospital, unable to deny his plight any longer, he began weeping and cried out, "I surrender, God. Take my life and do with me what you will. I am yours." There was no thunderous reply, but then for days after, what seemed to be miracles started happening. First he was informed by his physician that his liver was sound again. Then, for some inexplicable reason, his wife changed her mind and decided to give their marriage another chance. Then his colleagues welcomed him back to the clinic, and his patients, who had been offended by his off-putting behavior, began coming back. He did not feel he could attribute all these events to chance alone. "Some Power, not only all-knowing and controlling, but all-caring and compassionate as well, was operating. . . . You can tell me there's no personal God, but I'll tell you you're a blind fool."

While in the depths of despair, Bill Wilson, the cofounder of Alcoholics Anonymous, cried out, "If there is a God, let Him show Himself! I am ready to do anything, anything!" Suddenly the room lit up with a brilliant white light, and he felt an indescribable ecstasy. It seemed that he was on top of a mountain with a spirit wind blowing upon him. And then he realized that his bondage to alcohol was over, that he was a free man.[33]

Filled with a horrible self-loathing, the former Senator Harold E. Hughes picked up his 12-gauge, single-barrel Remington pump gun, slid three shells in the magazine, pumped one into the chamber, and climbed into the old-fashioned, claw-footed tub. With the shotgun resting on his stomach, he positioned the muzzle in his mouth toward his brain and placed his thumb on the trigger. A terrible sadness filled him. "Oh, God," he groaned, "I'm a failure, a drunk, a liar and a cheat. I'm lost and hopeless and want to die. Forgive me for doing this." Then he slid on the floor, sobbing convulsively. As he lay there exhausted, a strange peace settled deep within him, filling the emptiness. God was reaching down and touching him, a God who cared, a God who loved him. An intense joy burst within him. Slowly he rose to his knees and cried out in gratitude, "Whatever You ask me to do, Father, I will do it."[34]

It is important to note that not all revelatory experiences are of a religious nature, and the notion of a Higher Power need not always be equated with God. Atheists and agnostics may also experience personal awakenings. I have heard of AA members who could not believe in a personal God but who still had uplifting experiences. One, for instance, would pray to "To Whom It May Concern." Another addressed his prayers at a nearby light bulb during his regular AA meetings. And sometimes transcendental awakenings and profound, psychological illuminations happen and serve as the impetus for recovery. These cases serve as examples.

Randolph awoke one morning with a terrible hangover and shakes, and immediately began looking for a bottle of liquor he had put somewhere the night before. For some unknown reason, he happened to walk out onto the back deck of his country house and to look down in the woods. This was in February, and the woods were beautiful, with a sprinkling of light snow on the ground. Dressed only in his slippers, pajamas, and robe, he just started walking aimlessly for what seemed a long time. And then he sat down in the woods and cried. "I can remember that so well," he said. "I saw a rabbit, and then . . . a couple of rabbits. I saw a raccoon [and] birds. I just started looking at the beauty . . . and the serenity of the woods. I don't know how the feeling came over me, and I said, 'Who am I? What in the hell am I doing in this?' So I just sat there probably for an hour, trying to figure out who I am. Then I turned around, walked back to the house and told my wife I'll never drink again. But I'm going to do it one day at a time. . . . The next morning I got up and went to the woods again. I spent a couple of hours there and just wondered. I said, 'Well, I got one day in, I'm going to do another.' And I did that day. It was still hell. . . . And then pretty soon, in four or five days, I really got to feeling good. And, by God, I found out that life is beautiful."

After another drinking spree, Wesley was filled with remorse and shame. Another failure. The future looked bleak. Then one day at the treatment center, he informed his psychiatrist, Dr. Harry M. Tiebout, "I've got it!" The previous night he had experienced a psychological awakening, a sudden flash of understanding about himself as a person. For the first time, he could see all his rationalizations and defense reactions and just how selfish he had been. As the insights continued to unfold over the days ahead, he developed an inner peace and the certitude that he would never drink again.[35]

What do these vignettes reveal? As might be predicted from our earlier comments on bottom experiences, they suggest that glimpses of heaven are most likely to occur only after individuals have experienced the agonies of a personal hell. Transcendence or contact with a personal God seems possible only after a period of prolonged despair. Also, if these accounts are at all representative of others—and I believe they are—then it seems that these spiritual experiences tend to occur only when individuals are in a state of relative intoxication or have recently cut back or stopped all intake of alcohol, rendering them susceptible to the array of hallucinatory phenomena experienced during alcohol hallucinosis or delirium tremens. From my studies and search of the literature, I have yet to come across alcoholics who report profound experiences of this sort while mentally sober, emotionally stable, and physically sound. That is not to say that this situation cannot exist. I simply have not encountered it.

None of these latter observations detract from the authenticity of the spiritual experience and its potential impact on the individual, even though there is still no data on what percentage of alcoholics have this experience and continue to drink. As with religious conversion experiences,[36] which also often arise from the fertile soil of despair, the sudden awakenings of alcoholics may require an unstable emotional and physical state in order for them to experience the new-found sense of confidence, inspiration, and elation associated with the manifestation of a divine presence, and subsequently, the prospects of a radically altered value system.

In a sense, what transpires does seem near miraculous. Some alcoholics who have recovered this way even report the sudden and continued disappearance of the urge to drink, the craving that dominated their existence for so many years.[37] Nothing in medicine or science quite rivals this. No wonder so many individuals are convinced of the validity of their experience. All this makes William James' question just as pertinent now as when it was posed over fifty years ago: "Are conversions divine in causation (although divine in fruits) or like any other mechanism, high or low, of man's interior life?"[38] After weighing all the evidence, William James judged them divine. Perhaps if he had had the benefit of some of the newer insights offered by attribution theory, which attempts to explain how people attribute meaning to personal experiences or figure out what causes what, he might have hedged or concluded otherwise.[39]

It is not my intention to argue for or against the importance and necessity of a spiritual experience for the process of recovery, but rather to note that it is an authentic phenomenon that cannot be easily ignored. For those individuals who experience it, it is literally a "godsend." But these are only a small minority of the alcoholics seeking abstinence. Most, whether receptive to such an experience or not, must try to achieve sobriety by more secular or conventional means.

One practical matter also bears comment. Because profound changes occur—regardless of whether they are secular or religious in nature—is no guarantee that they will last. Backsliding, as with any religious conversion experience, is not only a possible but a highly probable occurrence.[40] Alcoholics can never rest on their laurels. Unless they continue to move forward, rather than waiting for a Higher Power to do it for them and chart every step of the way, they will likely find themselves slipping back.

The classical anecdote provided me by an English professor, an erstwhile member of AA, perhaps best captures the challenge for recovering alcoholics, regardless of whether they choose to pursue a religious or secular path to recovery.

Ulysses, the wisest and craftiest of all the Greek leaders, had volunteered to go on the most dangerous of all missions, to steal his way at night into the city of Troy, then to enter the Temple of Athena, steal the small golden statue of Athena called the palladium, and find his way back. The disappearance of the statue would make the Trojans think that their goddess had deserted them and that their city was doomed. Before he went on his dangerous mission, Ulysses got down on his knees and prayed to Athena. He reminded her that she had never before deserted him no matter what the danger involved. Since this was the most dangerous of all expeditions, he pleaded with her to accompany him into that strange, hostile city, to help him get that golden statue and bring it out. But then, after he had finished this prayer, he got up off his knees, looked up, and said, "But if you won't go with me, I'll go by myself!"

If God doesn't intervene, alcoholics will have to find a way of resisting temptations on their own.

7

Resisting Temptation

Larry, a thirty-five-year-old mechanic, felt that he had his problem licked after he was sentenced to prison. The urge to drink, which had dominated his existence before, miraculously left, leaving him confident about his ability to avoid alcohol in the future. During his confinement, he was given increasing responsibilities for managing parts of the prison farm. He was in charge of about 1,400 head of cattle. He gave all of them their shots, trimmed their hooves, delivered calves, and did whatever else was necessary. It was a seven-day-a-week, round-the-clock job. Though incarcerated, he had a considerable degree of freedom. At night, the guards would take him out of lock-up to check the cows. He was allowed to drive a jeep on his own. On weekends a girlfriend would visit. In fact, he admitted to finding his existence there almost idyllic. It was a totally controlled environment, devoid of any stress. He knew he couldn't drink, so he didn't. But that lasted only as long as he was confined. Paradoxically, the freedom he had gained from his addiction while in prison once again became a bondage to alcohol as soon as he was released and no longer had any constraints.

This story illustrates many of the mysteries about alcoholism, such as the importance of the physical setting for the appearance of craving, the matter of expectations about drinking, the availability of alcohol, and the therapeutic advantages of a structured lifestyle and keeping busy. What this story does not reveal is why this person so easily succumbed to his craving once released from incarceration. Hadn't he learned anything from his long period of abstinence? And what could he have done differently—what mental techniques could he have used—to avoid or reduce temptation?

From in-depth interviews with alcoholics, I have found that the nature of the subjective experience of craving varies widely. It can be an inchoate urge, almost without mental content or imagery, verbal thought, visual images, or a total experience including smell, taste, vision, and even physiological effects. To a large extent, the nature of the urge or compulsion depends on the predominant way each individual thinks.[1] Some people tend to think mostly in images, picturing in their minds the objects of their thoughts. They see themselves, for example, walking into a bar and ordering a drink. Others tend to think primarily in concepts, mentally speaking to themselves in partial sentences, phrases, or verbal shorthand. Others are visceral thinkers, experiencing a wish or intention in terms of bodily sensations or feelings. They almost smell or taste the imagined drink or feel its warmth spreading over their bodies. Others essentially think instinctively, short circuiting both imagery and cognitions, and are inclined to act without quite knowing why. When alcohol becomes readily available, they drink before they think. And others use different combinations of all these cognitive modes. The following are some actual examples.

Romance was Donna's weakness. She would picture herself across a candlelit table from a handsome man, music would be playing, and a bottle of wine added the perfect touch.

What Marjorie found irresistible was the image of herself sitting on the beach, her hair blowing in the wind, and sipping on a cold gin and tonic with a wedge of lime. Whenever she thought this, a voice in her head would utter, "This time it will be different," enticing her to drink.

John felt the urge in his belly, a trembling, like fear, that only a drink could quell.

Matt loved the taste of a martini. At times he would find himself overcome by the urge. In his mind, he imagined himself savoring the distinctive flavor, swallowing some down, then feeling the warmth radiate through his belly and spread throughout his body.

Jane experienced neither images nor words. Instead, she imagined a "wonderful feeling," a type of contentment that alcohol would bring.

For Quinn, the irresistible urge was a feeling, without pictures or words. It was like taking a bite of a juicy steak that had been prepared especially for you and then having it suddenly taken away.

When Richard craved a drink, he could almost feel its effects as he imagined its taste and visualized himself relaxed and talking wittily.

Jason, too, thought in terms of images and sensations. Particularly on hot days or on appropriate occasions, he could almost taste the cold beer washing down his throat and feel the pleasant distention in his belly and the tingling buzz in his brain. At times, he also would use little buzz words or phrases, like "It sure would taste good" or "Satisfy your thirst" to capture the fleeting thought.

The imagery and almost hallucinatory sensations were also vivid for Greg. After doing some physical work, he frequently would think of a beer. He almost could feel the coldness, experience the wetness of the can in his hand, see the rivulets of moisture running down the side, and then imagine the taste. Then he would mentally compare the taste of different beers, and imagine gulping one or the other down.

These are all typical accounts from alcoholics. Common to all is the claim that if they fondle these thoughts, if they dwell on the taste and effects of the wine, liquor, or beer, the compulsion to drink will become unbearable.

How do the successful alcoholics manage to overcome these temptations? The first thing they must do is to recognize just how clever, persistent, and seductive their minds can be at providing them with justifications to drink.[2] The images they conjure up serve this purpose well. Almost any reasons they use, no matter how convincing, that make liquor more easily accessible, that make "accidental" slips more likely, or that justify their taking only a drink or two, are only ways of conning themselves so that they can experience a guiltless drunk. Bad luck doesn't just happen. Alcoholics subconsciously work at making it happen. The enemy is not the bottle, but their own perverted logic that puts alcohol within easy reach. Slippery thinking, not an evil genie, is responsible for a slip.

The next general axiom is that alcoholics must never relax their mental guards and regard themselves as immune to temptation. The urge to drink may lay dormant in alcoholics for years, lulling them into a state of complacency and false confidence, just waiting for the right moment to arrive, for the proper state of mind, and then strike them with full force. The preconditions for this type of recrudescence are many. Recognizing the absence of any real cure and that the latent vulnerability to relapse is always there, recovered alcoholics must remain on their guard for early warning signals of dangerous thoughts that are apt to arise whenever their resistance is low. Eternal vigilance is not just the price of freedom; it is also the price for sustained sobriety.

Despite these precautions, individuals may still find themselves wrestling with the urge and progressively weakening during the struggle. Most alcoholics eventually succumb, but some manage to survive. The survivors tend to be those who, through either trial-and-error learning on their own or instruction from others, have discovered a variety of mental techniques for effectively coping with temptation. In many ways, these techniques are similar to those commonly rec-

ommended to deal with obsessional thoughts, compulsions, or phobias.

One of the most common methods for coping with obsessions and
temptations—and probably the least efficient—involves counterpoint
thinking. It is this type of thought, the clashing of opposites, which
underlies painful attempts by alcoholics to wrestle with a craving. It is
successful application of this thought, the mobilization of mental energy
against an unwanted thought or impulse, which supposedly underlies
the expression of willpower. Unfortunately, this method of coping,
while helpful for some, has too many inherent limitations to be considered as very effective. It is usually the first approach adopted by
alcoholics during the initial phases of recovery, during the time when
they are constantly beset by cravings. It is also, for most, the first
approach to fail.

In Irma's case, for example, the constant bickering in her head would
almost drive her crazy. It was like four shrill voices saying the same
things over and over. One voice would taunt her with "Go ahead,
take a drink. One won't hurt." Another would reply, sarcastically, "Sure,
you'll only take one, but you're going to have a blackout and not know
what you're doing. You're going to get big and fat again, and everybody's going to hate you." Then another voice would answer, "So what?
I don't care." And then another voice would try to shout the others
down with, "You all shut up. I need it. You haven't been through
what I've been through." And this would go on without end.

What the individual battling a persistent and powerful craving must
realize—and this is one of the most fundamental concepts for dealing
with any obsession, urge, compulsion, or idée fixe—is that fighting
the craving head-on only fuels it. It paradoxically gives power and
import to what is only an insubstantial thought or inclination. Thinking a thought does not mean acting on it. But by engaging in this
counterpoint thinking, by allowing this battleground to exist in the
mind, the individual, in essence, is indicating that the unwanted thought
or impulse, if uncontested, can force him to do something he really
does not want to do. The I-want-a-drink versus You-can't-have-a-drink
battle becomes a dialectic without resolution—thesis, antithesis, but

no immediate synthesis. Fighting an obsession is like engaging in a marital squabble in one's brain. No wonder so many wilt under the endless harangues, relinquish the struggle, and capitulate to the urge. It simply is not worth the effort to resist. But the problem is that the fight was rigged, the outcome preordained. What would happen if the individual never challenged the obsessional idea in the first place, if he stepped out of the ring? The thought would be left there alone, to shadowbox ridiculously with itself, doing no harm.[3] There can be no real defeat if the individual chooses not to fight, if he chooses simply to ignore the bullying thought. Or, as AA proponents might claim, if he must fight, let his God get in the ring in his stead.

The reason why most New Year's resolutions do not last also sheds some light on the unproductive and mostly counterproductive nature of counterpoint thinking. The problem with a New Year's resolution,[4] in which an individual promises to give up a certain behavior, like smoking, sweets, or drinking, is that it is difficult to picture oneself *not doing* something. The picture of not doing triggers an image of doing what the individual is resolved not to do, thereby compounding the problem. The reason that it is hard is because the words—which are really symbols for objects—tend to trigger the corresponding images, but individuals cannot visualize "not." No wonder then that so many alcoholics become frustrated when fighting the image of themselves consuming a beer or cocktail by trying to conjure the image of themselves not consuming a beer or cocktail. This also helps explain the sense of helplessness, lack of willpower, and loss of control when the very thought or image they are fighting not only continues to persist but actually grows in intensity, defying every self-injunction to disappear.

The answer to this dilemma then is *not* to visualize what one is *not* going to do, but to picture a substitute or opposing behavior that one *is* going to do. To begin a new habit, the individual should have a behavioral image of himself engaging in the new behavior, not just on occasion but every time the unwanted habit pops into mind. New habits are formed by repetition. The more frequently the individual conjures up the new behavior, the more automatic and natural it becomes. In the case of the alcoholic, for example, this might mean that instead of picturing himself *not drinking* alcohol when the craving occurs, he would picture himself drinking a soft drink, tea, or juice, and imagine himself enjoying its taste. If he then acted on this new image on occasion and actually drank the nonalcoholic beverage, he would eventually give more power to this image and weaken the old one.

There are many other ways to cope with a compelling urge. Distraction appears to be another common technique.[5] With this technique, individuals attempt to shift the focus of their attentions away from the unwanted or taboo thought or intention onto another activity. Typically with distraction, the individual tries to keep busy with arduous physical activity, social interactions, mental pursuits, or with any variety of tasks that will serve as diversions from the obsessive inclination to drink. This is the basis for the claim by William James that the cure for dipsomania is religiomania, and the oft-cited notion that the remedy for alcoholism is workaholism. In an effort to avoid the clutches of the compulsive thought, some individuals try to keep constantly busy at work, get involved in all kinds of civic activities, become zealous members of their churches, or become physical fitness aficionados. A lawyer I knew, for example, handled his persistent urges to drink after he decided to quit by studying to become an accountant while still carrying on his full-time practice. A physician began spending more and more time with his patients and the hospital staff. Another individual, an engineer, became a marathon runner and health food enthusiast, devoting all his spare time to mastering these pursuits. And then, of course, there are the thousands upon thousands of recently dry alcoholics who try to follow the AA dictum of 90/90—attending at least ninety meetings in the first ninety days— as well as spend whatever time they have left in other constructive pursuits. If they keep busy not drinking, then they cannot keep busy drinking.

Turning to someone for help can also be a form of distraction, a way of getting one's mind off troublesome thoughts. For example, one individual, the president of a coal company, had been dry for only about six weeks when he experienced one of his worst compulsions to drink. He was in his office at the time, emptying his desk, when he came across a half-pint of bourbon and three beers in one of the drawers. "I was too scared to leave, too scared to stay there," he told me years later. "I knew that if I started home I had to drive by three bars that I used to drink in. So I called my sponsor in AA, who was never home during the day, but that day he was home, and as soon as I heard the sound of his voice, the craving went away. I was still shaky. I asked if I could come over, and he said 'Yeah.' So I went over. We didn't even discuss my compulsion to drink, but he knew why I was there. We talked about the weather and this and that and then went out and had a bowl of ice cream. Then I went on home."

Though distraction has its merits, it also possesses some of the same, inherent failings of counterpoint thinking. In a sense, what the individual really is doing is running away from his problem, doing everything in his power to avoid thinking about the unthinkable. One alcoholic I interviewed prayed fervently, day and night, for strength to resist the urge. Another would meditate, argue with himself, eat honey, take a hot bath, jog, start making phone calls to friends, try to read, or smoke incessantly when he was by himself in an effort to keep the tempting thought from taking possession of his mind. While deliberate distraction may successfully keep the urge at bay, it unfortunately reinforces the notion that there is something to be frightened about. This, in turn, paradoxically gives even greater power to the unwanted thought. With this approach, the problem is that no matter where the individual flees, no matter how much he manages to distract himself, the monster will always be in pursuit. In time, the intensity of the urge and the frequency of its appearance may diminish, but embers of the urge will continue to remain for long periods of time on the back burner of his mind, ready to flare up anew, mainly because the urge has never been extinguished.

There is a dream I like to tell to my patients that has some bearing on the problem. In this dream, a green, scaly, fire-breathing monster with long talons is chasing a frightened young lady who, just as she seems about to escape, trips over a rock and falls to the ground. With the monster poised over her, she cries out in terror, "What are you going to do to me?" A look of puzzlement comes over the monster's face as he says, "I don't know lady, it's your dream."

The moral of this story is that there is no need to fear the monster. Whatever the monster eventually does or does not do is entirely up to the individual.

The reason that this urge retains so much power, the reason alcoholics try to argue with it or escape from it, is that they implicitly believe in the omnipotence of thought—the belief that because a thought exists, it must be a true reflection of how they feel and, as such, will likely be acted upon unless they try to thwart it. If instead, they can come to regard the urge as a foreign, aberrant happening or irritant in their brain, then they can feel freer to ignore it or, like the frightened woman in the dream, shape the outcome to suit their own ends. The urge can scare them only if they let it.[6]

One favorite technique for manipulating the urge and the self-justifications that surround it is to "play the script out" or "think through the drink," not just fix on the image of pleasure associated with alcohol, but to go and play the mental reel out to the end and include all the detrimental consequences that occur as drinking continues—the drunkenness, the failures, the humiliation, and the sickness. In my study of alcoholics who made spontaneous recoveries, almost half still experienced automatic positive associations of drinking, but they were able to blunt the effect of these positive associations by continuing their trains of thought to the bitter end, thereby negating the original urge to drink. Here are some personal descriptions of the techniques they used.

"I think, 'Wouldn't it be nice to have Scotch?' and can just about taste it. Then I see myself about four Scotches later . . . being sloppy. Or I can remember my personality changing. . . . You know, not sitting there having a pleasant Scotch, but being drunk."

"A beer still smells good. . . . The smell of alcohol before somebody digests it doesn't bother me. The smell afterward is . . . close to being nauseating. The smell of foul food, rotten meat, whatever."

"At first I had the craving . . . when I craved a drink I would almost see, feel its effects, and see myself relaxed, talking wittily. . . . I dealt with it by straight Pavlovian conditioning. Every time I had the urge to drink, I would immediately think of being sick, vomiting, shakes, being miserable. At first this was conscious, but no longer so."

A female artist and writer, who claimed to be mainly an image thinker, informed me of a most clever way she hit upon to control her impulse to drink. For her, a romantic image of being at a party, the center of attention, or sipping a drink at dinner with a male companion served as a powerful inducement to drink. With her commitment to sobriety, she hit upon a mental technique of putting captions under these images or superimposing one image on another. For example, if she imagined herself talking to a man while drinking at a party, she might put a caption beneath it of "Drunken Suburban Housewife," certain that her inevitable intoxication would drive the man off. Or she would superimpose an image of herself drinking coffee over the one of her with a martini, with a new caption of a man asking "Can I Call You Up Tomorrow?" From past experience, she knew that she was much more attractive to men when sober than when inebriated.

This technique also has elements of another technique, known as "substitution," in which an individual selectively substitutes the image of a more suitable object or situation for a less suitable one. A glass of wine becomes a cup of coffee, a cold beer becomes a soft drink, a bar becomes a comfortable living room, and the imagined wittiness associated with drink becomes the blabbering inanity associated with drunkenness.[7]

What this technique illustrates so well is the individual's ability to manipulate symbols to his own ends, to direct his own scenario. As long as he does not confront the symbol (e.g., the urge or the rationalization to drink) head-on by trying to argue with it or eliminate it directly from his thoughts, he has the capacity to mold and shape it so that it will have a negative or opposite effect. The technique is analogous to what a boxer does when he rides with a punch or deflects the momentum of a blow, or the type of block an offensive lineman might employ to alter the course of a hard-charging, defensive tackle on the opposing football team. The aim is to use the force or momentum of an adversary to one's own advantage, an approach that has been developed to a scientific degree in such martial arts as jujitsu.

Allied with this technique is the one of immediate negative conditioning, whereby the initial thoughts about drink are linked with very aversive consequences. These consequences can be represented in terms of physical feelings, undesirable emotions, or upsetting situations. A substantial proportion of recovered alcoholics utilize these techniques.[8] Here are some sample comments.

"Whenever the urge hits me," one individual claimed, "I think back to the worst time, such as waking up in jail and the blackouts and the worrying about whether I'd hit somebody with my car and the time that the fellow brought me home, where my wife and two boys saw me. It was the first time my boys had ever seen me like that."

"The episode is there with me all the time, lurking in the background," another person said. "I keep remembering what alcohol did to me in the past. I sort of picture a gun up against my head, and I'm ready to pull the trigger and hear a thud."

"I can take a bottle of whiskey and uncork it, and the smell makes me nauseated."

With most alcoholics who employ this technique, the negative association becomes automatic and bypasses or short-circuits the stage of positive images and connotations involved with alcohol. The thought of a drink becomes equivalent with an unpleasant experience. In some cases, when this association is not automatic, individuals learn how to deliberately call up these negative associations at the very first inkling of any desire for alcohol, thereby keeping any stronger urge from developing.

It is of interest to note that this way of dealing with cravings and rationalizations bears striking similarities to a more formal treatment technique, known as covert sensitization, which has been widely employed for a variety of habit disorders.[9] In the case of alcoholism or problem drinking, individuals are trained either through hypnosis or mental imaging techniques to associate an unpleasant sensation, like nausea, with the notion or image of reaching for a drink. These pairings are done numerous times, until the experiences of nausea after the image of drinking becomes more automatic and presumably "conditioned." Despite the theoretical rationale for these procedures, the clinical results have been unimpressive in producing long-term abstinence. The reasons are unclear, but they probably pertain to the strength of the conditioned response, the degree of motivation in individuals, and the matter of whether the right aversive response has been paired with the right drink-related stimulus in the right person at the right time. What is humiliating or aversive for one alcoholic may not be for another, just as what may be a powerful drinking cue for one may not be for another. The situation is complicated by the observation that drinking cues for an individual tend to change over time or be different in different drinking situations. For many alcoholics, conditioned, standard experiences of nausea in response to the thought of a drink only prove to be weak deterrents to the attractive alternative of intoxication.[10] It is much more likely that the conditioned experiences will influence alcoholics more effectively if they involve painful events that hold special individual significance.

Aside from this mental aversive conditioning, alcoholics may also use the related techniques of thought-ignoring and thought-stopping, techniques proven potentially beneficial in the treatment of many forms of obsessions.[11] With thought-ignoring, the individual treats the disturbing thoughts as foreign elements and attempts to ignore them, as with pain, or simply keeps reminding himself, over and over, not to pay attention, or saying to himself, "ignore, ignore." This approach can be regarded as the mental equivalent of a parent deliberately ignoring the temper tantrum of a child so that the behavior will not be rewarded.

With thought-stopping, the individual tries to check the thought or disturbing image at its very first signs of appearance, before the momentum of the thought processes has a chance to build, by cutting into the flow of thoughts with an imagined barrier or a subvocal, personal admonition, "Stop." The expectation is that the more often individuals voluntarily practice and successfully apply these techniques, the more automatic these techniques will become, and the less that the unwanted thoughts occur, the more likely they will disappear.[12]

The thought-stopping technique can also be applied to dangerous or slippery thinking. One individual, for example, had ingeniously set himself up for a slip by consistently mishearing what was said. At the time the incident happened, he had been sober for about two months. He and two other business executives were meeting for lunch. The others ordered bloody marys, then, when the waitress asked what he wanted, he said he would also take one, thinking that they had ordered iced tea because, so he claimed, he had heard only the last syllable. So the waitress brought three bloody marys and put one in front of him. "I looked at it, and it looked at me," he informed me much later. "And I thought 'what are these guys going to think of me if I don't drink it, particularly after she brought it. What the hell, you're 200 miles from home and you probably could drink one anyway. And if you get drunk, no one is going to know.' And then something in the back of my head said, 'Quit bullshitting yourself!' " So he forced himself to stop thinking about drinking, sat back in his chair, and told the waitress to bring him some coffee instead.

Postponement appears to be another effective tactic. The individual derives consolation from telling himself that he doesn't need to battle his craving for a lifetime but only for the next twenty-four hours. Or if that still seems like an infinity, then only for twelve hours. Or if that seems too long, then only for six hours. And on and on, even if it is only for an hour or fifteen minutes. Surely, the individual tells himself, he can hold out that long. Then if he makes it that long, he assures himself that he can endure another such interval. Then another, and another. And then maybe, sooner or later, the urge will let up, and he can have some peace again.

Despite this diversity of mental coping techniques, all tend to have one common goal. They are designed to hold the unwanted impulses at bay—to stop them in their tracks before a juggernaut momentum can build—so that more constructive thought can take place. As to how these techniques work, only speculation can be offered. In my opinion, though, information theory represents the best way of explaining what happens.

According to information theory, a message or signal can be jammed by transmitting another message or signal that is 180 degrees out of phase. This is how radio or television signals are jammed. This censorship, neutralization, or jamming can also take place in the mind when one message is simultaneously substituted for another, as when individuals tell themselves "I hate alcohol" at the very moment "I want alcohol" comes to mind. Or, instead of imagining the pleasant smell and taste of alcohol, they imagine the shakes, a bad hangover, or gastritis, instead of a nice "buzz," they imagine themselves as tearful, depressed, or guilty, instead of wittiness, they imagine blackouts and embarrassment, instead of picturing themselves with other "charming" people who are drinking, they envision the disappointment of family and friends.

Two important features of this jamming model need to be noted. First, it is not necessary for the competing message to be exactly 180 degrees out of phase with or diametrically opposite to the original message to make it unrecognizable. Almost any different or contradictory messages or signals will do. This means that individuals should be able to combat unwanted images or thoughts, not just with opposite ones but also with any that are incompatible, such as with thought substitution, thought manipulation, thought-stopping, and distraction. Second, the most effective way to jam a message or signal is to have the competing signal occur as nearly simultaneously with the original signal as possible. The later the onset of the competing message or signal, the less its ability to jam or neutralize the original one. That is probably because it allows some part of the original message to get through. The part that gets through then stimulates a whole train of powerful mental associations and conditioned responses, such as those associated with drinking. The key to preventing this from happening, at least theoretically, is to obliterate or jam the disturbing message or urge before it becomes clearly recognizable and established—to stop it before it triggers a train of unwanted thoughts and responses—and not try to counter it after the fact, as happens with counterpoint thinking. However, countering a message after it is transmitted is better than nothing at all, for it at least suggests an alternative response, even though it tends to take the form of frustrating, back-and-forth obsessional thought. Most individuals who have achieved a stable recovery not only become proficient at dispelling these unwanted messages or urges from their minds, but also learn to shield themselves against them even before they arise.

What is so remarkable about the techniques used by recovering alcoholics to resist craving is that they are similar to those used by Buddhist monks centuries ago, as we shall see, to control their unwanted, intrusive thoughts during meditation,[13] suggesting both a timelessness and universality for those mind-control strategies.

In early Buddhism, monks and lay disciples learned how to keep unwanted, unwholesome thoughts from entering their mind so that their meditative efforts would not be disturbed. The Vitakkasanthana Sutta of the *Majjhima Nikaya*, compiled shortly after Buddha's death in the 5th century B.C. and its commentary, found in the text *Papancasudani*, recommend a series of techniques to accomplish this, each to be tried hierarchically if the preceding one fails. As an initial strategy for controlling intrusive thoughts, these writings advise the novitiate to *switch to an opposite or incompatible thought*. This is accomplished by reflecting on an object or thought that is contrary or diametrically opposed to the unwanted thought. Lustful sentiments, for example, should be offset by those associated with purity, greed by charity, or hatred by love. If that doesn't work, the individual should *ponder on the harmful consequences, dangers, or disadvantages of the thought*. The price of immediate gratification may be long-term suffering. If that strategy also fails, the individual should *ignore the offending thought*. This can be accomplished by turning the attention elsewhere or distracting oneself by memorizing and recalling doctrinal passages, concentrating on actual objects or activities, like reciting a mantra, or engaging in an unrelated physical activity, like mending one's clothes. Beyond that, the individual should *reflect on the removal or stopping of the causes* of the disturbing thought. This technique can be understood by the analogy with a man walking quickly who asks himself why he is doing that and, seeing no reason to do so, slows down. Then he reflects on his walking and stops. Then he reflects on his standing and sits down. And then he reflects on his sitting and lies down. Finally, if none of these various techniques work, the individual should *control the thoughts with a forceful effort*, with clenched teeth if necessary. The analogy is given of a strong man holding and restraining a weaker man, or of someone using one part of his mind to dominate the other.

But beyond the application of these techniques, whether by recovered alcoholics, Buddhists, or anyone struggling with unwanted thoughts or compulsions, there is one general condition necessary for continued success, and that is the issue of personal resolve or commitment. Sobriety is unlikely to occur if the individual undertakes it in a halfhearted, perfunctory, ambivalent way. It involves total effort.

As long as the individual remains committed to sobriety, as long as he continues to regard the consequences of drinking as aversive, as long as he entertains no ifs, ands, or buts about not returning to drink, he will fare far better in controlling the urge to drink than if the option to drink is still present. With this type of resolve, stimuli that usually were capable of eliciting craving for alcohol not only may no longer do so but now may make the thought of a drink repugnant. But let the individual let down his guard, admit to the possibility of moderate drinking, and he opens his mental floodgate to all sorts of rationalizations and powerful urges.[14] That is one of the problems with alcoholics who aspire to become social drinkers. By always having to decide whether to drink or not to drink and, if they drink, how much, these individuals will be forced to marshall all their energies and coping mechanisms to deal not only with an ever-present craving but also with the chicanery and deception from their own minds.

The reason why many alcoholics relapse is that they eventually relax their guards and lose the intensity of their convictions about the importance of sobriety. This is most likely to happen in situations that have been risky for them in the past, times when they were most drawn to alcohol. That's when the "bad luck" starts and the temptations become overwhelming. It is not that sobriety is no longer important, it is that sobriety is not *all-important*. What is required is a "totalitarian" frame of mind, one that will tolerate no compromises or exceptions. Without that absolute commitment, that constant vigilance against rationalizations, individuals will soon find themselves succumbing to temptation. With a laissez-faire mentality in which competing intentions can vie for dominance and the most expedient are likely to win out, urges and cravings for alcohol will continue to play havoc.

Perhaps I can illustrate the importance of maintaining an absolutist or totalitarian frame of mind, along with certain uncompromising attitudes and images, by describing my own experience with another bad habit, overeating. A while back, tired of hearing the gentle teasing by family and friends about my increasing weight and disgusted with myself about my growing waist and "shrinking" trousers which no longer fit, I decided that enough was enough; I had to start gaining control of my eating again and bring my weight down. Armed with a newfound confidence and zeal, I launched into a strict diet and exercise program. Within six weeks I lost seventeen pounds, dropping from 182 to 165, but the remarkable thing was that I had absolutely no difficulty. Imbued with a sense of nobility and virtue, I turned down offers of my favorite desserts, ate only half-portions of favorite, but fattening, ethnic foods, avoided evening snacks, and experienced no particular hunger or distress. The weight reduction program was easy—until my wife told me that I had lost enough weight and should relax my stringent diet and shift to more moderate eating habits, and I agreed! From that moment on, as I had adopted a different frame of mind, one that permitted an occasional indulgence and a more balanced approach to dieting, my life became much more complicated as I tried to keep my calorie intake within "reasonable" limits, one that would sustain me at this lower weight. Temptations that did not exist before now came to dominate me. I began getting more obsessed with food.

Television commercials of juicy hamburgers, tender shrimps, roast beef sandwiches with melted cheese and crispy potato chips caused me anguish. I spent the evenings struggling against going to the food cabinet and refrigerator where I knew boxes of delicious crackers, cookies, nuts, raisins, packed meats, and cheeses sat, just waiting to be consumed. At first I resisted this temptation, eating only "nutritious" and "low-calorie" fruits, vegetables, and other "health" snacks, but soon I began taking little nibbles of the assorted delicacies that now seemed more noticeable about the house or just tiny bits of second portions at meals, not enough to make a difference, so I told myself, and I began avoiding the scale, even though in my own mind I was still on a strict weight maintenance program. Then finally, one day I screwed up my courage, and I was shocked to find that I had regained seven pounds. My immediate reaction was to assume that the scale was in error, since I still believed that I was dieting diligently, but I had no desire to recheck my weight on another scale. Deep down, I knew what was happening, but consciously, didn't want to know that I knew. All of my hard-won gains were about to be lost, and I just couldn't seem to help it. My mind became even more cunning, providing me with a whole slew of ingenious excuses for indulging without restraint, even though I really wanted to look good and remain slender. All this lasted until I could no longer deny what was happening. That realization forced me to take action. So I again grabbed myself by the scruff of my psychological neck, so to speak, and recommitted myself to a strict dieting program. With this renewed commitment, I was once again in control.

It was this personal experience with dieting that provided me with the insight into the seemingly unfathomable changes occurring in alcoholics. It showed me how individuals could shift suddenly from a state of obsessional preoccupation with alcohol, constantly besieged with urges and cravings, to one of complete aversion to alcohol and relative immunity to temptation once they had made a firm commitment to quit drinking. Their teetotalling, absolutist frame of mind, along with its associated attitudes, simply closed off the option to drink, and in essence, made the experience of craving superfluous. If drinking is no longer an option, then why dwell on its delights? At the same time, it explained how other individuals, after years of successful sobriety, should find themselves once again overcome by temptations, struggling against the desire to succumb while allowing their minds to engage in slippery thinking and to maneuver them into situations where it becomes inevitable that they fail. No sooner do alcoholics let down their guards and, for whatever reasons, adopt less stringent expectations about drinking, then they find themselves becoming more and more obsessed with drinking. Paradoxically, they may still want to remain sober!

But it is an easy matter to realize the dangers of letting one's guard down and another matter to avoid doing so. It requires the expenditure of a lot of mental energy to remain wary and aware, constantly on the alert. Yet alcoholics have no choice. If they are going to preserve their hard-earned sobriety, then the sentry in their minds can never leave its post. For that to happen, individuals must remain "psyched up," providing themselves with regular pep rallies within their brains that can help to increase their immunity, as booster vaccinations do with disease, against relapse. They can accomplish this by taking pride every day in their continued sobriety, maintaining an active interest in fellow alcoholics or the field of alcoholism, and providing themselves with frequent reminders about their vulnerability to slip and the dangers of certain types of thought. If individuals have difficulty doing this on their own, they can turn to an organization such as Alcoholics Anonymous, which provides opportunities for this to occur during meetings, readings, and discussion groups.

In time, perhaps, as their urges extinguish, as confidence in their coping techniques increases, and as their attitudes about abstinence crystallize and harden into a firm resolve, individuals will be able to maintain their vigilance with less effort. By then, they should have developed certain ways of thinking that are antithetical to those fostering drinking. These new ways of thinking, and the attitudes associated with them, represent their first line of defense against relapse.

8

Sobriety Scripts

When he was thirty-seven years old, Eugene O'Neill underwent a six-week, experimental psychoanalysis with Gilbert V. Hamilton.[1] Though past treatment for his alcoholism had not helped, this time it was remarkably successful. Except for a couple of insignificant lapses, he remained sober until his death twenty-eight years later, banning alcohol from his house and, with his third wife, going into seclusion and shunning his former drinking friends. Unfortunately, the details of this remarkable transformation remain a mystery. Little is known about what was said, done, and thought during this relatively brief, therapeutic interlude and what happened afterwards that allowed such radical and lasting changes to occur.

Though we can never know with certainty what took place within this famous playwright's mind, we can gain glimpses of what he must have thought from the personal accounts of others who have achieved equally impressive results on their own initiative or after exposure to other sources of help, such as Alcoholics Anonymous, the church, or various therapeutic programs.

One of the main ways that individuals control what they do is by telling themselves what to do and what not to do in their minds. Often, these mental instructions are in the form of brief motivational sayings, like "Say 'No' " or "Don't drink," that have the capacity to quell temptation and bolster their spirits temporarily. The goal of all these aphorisms, slogans, one-liners, sayings, cue words, commands, or inspirational messages is essentially the same. It is to immunize individuals for a while against the craving for alcohol by reaffirming their basic desire to remain sober.

That this kind of inner speech or self-talk can be helpful is attested to by the great store that members of Alcoholics Anonymous place in certain buzzwords or sayings. These sayings, which are laden with implicit meaning, serve to keep attitudes and thoughts within proper bounds and to map out guidelines for appropriate behavior. The more catchy and pithy the sayings, the easier they are to remember. FIRST THINGS FIRST. EASY DOES IT. LET GO AND LET GOD. THIS TOO WILL PASS. ONE DAY AT A TIME. IT CAN BE DONE. A PERSON WITH ONE EYE ON YESTERDAY AND THE OTHER ON TOMORROW IS COCKEYED. THE GRATEFUL ALCOHOLIC IS A SOBER ALCOHOLIC. AVOID THE FIRST DRINK. BE POSITIVE. And so on. All these simplistic prescriptions for living are not only antithetical to the alcoholic way of life but are consistent with the rationality, maturity, and other positive qualities attributed to the sober state of mind.[2]

From my talks with alcoholics, I have become more and more convinced about the sustaining power and inspirational effect of such words. Like personal mantras, they can focus attention and serve as guides for action. Many individuals do not think in long, sequential sentences laden with abstract meanings and connotations, particularly when they are under duress. In those instances they are apt to cling to short, telegraphic statements that convey the necessary message in the most pointed and palatable form. And just because the statements smack of being platitudes makes them no less useful or true. Many individuals, in fact, derive a certain comfort in knowing that they are relying on "tried-and-true" instructions that presumably have proven helpful to countless others before.

In my own case, I first became impressed with the value of buzzwords and slogans some years ago in an entirely different context when I became lost in a forest after our canoe capsized and broke and the prospects of rescue seemed bleak. As I trekked though the wilderness with my companion, with no survival gear to rely on, I found myself thinking of hackneyed mental phrases or inspirational sayings, like "Keep cool," "A little pain won't hurt," "You've got to press on," or "There's nothing to fear but fear itself," in order to bolster my flagging spirits and maintain a more positive frame of mind. Surprisingly, I found these reminders very reassuring and helpful, although I was later annoyed with myself for not having been more original.

But this is not the only kind of thought alcoholics use. They also have available their reasoning processes—their thoughts, fantasies, images, and ideas—to shape and mold their experiences so that they conform more to their basic expectations for sobriety. Telegraphic statements may not be enough. Individuals may need to spell out the rationale for what they do in much more detail. In a sense, these thought processes represent the antithesis of those already described for "dry drunks" or "stinking thinking." They, too, tend to conform to standard "scripts," but the arguments or polemics are used to create a mental climate hostile to drinking and favorable to abstinence. As long as individuals think these thoughts and respond appropriately, they cannot, by definition, relapse.[3]

Let's examine how this applies to recovering alcoholics. An individual I interviewed gave this personal account of what happened to him one evening and how he managed not to slip.

"I was walking home from an AA conference that was held downtown that weekend. And there was a dance there. They had some sort of band. And I was there alone. And I sat there, feeling indifferent. Usually I'm eager to dance. I want to dance with everybody. But I didn't feel like that then. I felt a little anxious. I felt loneliness. So I started walking home. And thoughts of drinking came into my mind. I could stop and buy a six-pack on my way home and just drink it and forget everything. Just relax and blot out all this nonsense from my mind and go to sleep. And immediately, I thought to myself, 'Now how would you feel the next day? Think of the guilt you would experience. It's ridiculous. And you wouldn't stop there. You would want to go out and buy some more. And then the next night, and the next night, and the next night, and then you'd be ashamed to talk to people at work even if they didn't know that you'd drank. You would be ashamed of yourself. Then you would feel the same sense of hopeless desperation that had left you so recently. You might not be able to stop in a couple of weeks and go back to AA and pick up where you left off.' I was thinking these things on the way home. Then I went past two taverns that I used to frequent. And I thought as I went past the first one, 'Now there's nothing in there that you haven't already experienced two dozen times. There's no happiness or pleasure in there that you haven't looked for many times before. Again, and again, and again. So just walk past. There's nothing in there except dreary old tables with beer stains. You made the decision to stay sober, now stick to it.' And so I went past. And then I went past the other tavern where I used to hang out. Then I came to a couple of liquor stores, and the thought occurred to me to buy a six-pack. Or two six-packs. One six-pack would hardly produce a nice state of intoxication. But then I told myself that these thoughts were just a passing fancy, that they had come into my mind countless times before. They had no reflection in reality. They didn't make any sense. Just let them go in one ear and go out the other. Then I got to the last liquor store on my way home. It was open and the lights were bright and there were people standing behind the cash register. But I just walked past. I don't remember exactly what I said to myself, but it was something to the effect of, 'Well, you could always go to your own funeral tonight,' something exaggerated like that. But maybe that wasn't an exaggeration. Anyway, it was helpful for me to hear that, to hear myself saying that. Then I just sort of chuckled to myself and crossed the street and went on home. And when I woke up the next morning, it wasn't just an overwhelming sense of relief I felt. I felt a sense of well-being. I had made it by the skin of my teeth."

And he should have felt good about what happened. He had run the gauntlet and come out unscathed.

In this one anecdote, we find elements of at least four different cognitive themes being evoked. This man reminded himself about the negative consequences of drinking. He convinced himself that he couldn't handle even one drink. He noted that his former decision to quit drinking had been based on good reasons. And he reminded himself about the benefits of being sober, how he had escaped from that terrible loneliness. He did everything but pray.

The thought processes and images displayed during this man's long walk home are not so different from those most recovering alcoholics use almost every day of their lives when dealing with their periodic compulsions to drink or to strengthen their basic resolve for sobriety. These same four themes, with the addition of prayerful thought, tend to be reflected after a while in the implicit and often subconscious attitudes of individuals, which become explicit only when they resort to self-talk or are called upon to explain the way they feel. Each of these themes is represented by a separate mental script, along with its associated images, which tends to be remarkably similar for the individuals who use it. Some individuals, like the person above, employ elements from virtually all scripts to cope with difficult or risky situations, but others may rely on only one main script or a combination of several. Somehow or other, if alcoholics are to remain sober, they must learn to think in these terms.[4]

In my in-depth interviews with alcoholics in various stages of recovery, I have had the opportunity to document the actual thought processes that sustained them in moments of temptation or as a prophylaxis against relapse.[5] The excerpts presented represent their introspective recollections of what went on in their minds at these times. Because these thought processes and their associated images are so important for recovery, I have decided to present them in detail.

One of the commonest means individuals use to prevent or curb the desire to drink involves reminding themselves of all the bad effects associated with drinking. This is what transpires with the *negative consequences script*. Recovered alcoholics frequently invoke bad memories associated with prior drinking bouts, times when they were embarrassed, ashamed, humiliated, guilty, out of control, physically ill, or acting contrary to their basic moral standards. These reminders not only can be used prophylactically to alter basic attitudes toward alcohol but also as specific techniques, as described in the last chapter, to counteract craving. By constantly invoking these painful memories, individuals eliminate the option of drinking as a solution to life's problems.

Here are examples of what alcoholics actually tell themselves or think.

I see myself with a glass in my hand—mouth wide open—screaming at the kids.

What about the wrecked cars and the jail and the O.D.? Doesn't sound so great, does it?

I see myself throughout dinner, making a real fool of myself in front of all those people.

I see myself starting to drink—the pint of rum—sitting out and watching the sunset. Getting some wine—then having no idea of where I went—if I murdered anybody—or what I did that evening. Next morning the awful hangover. I detested myself—it was the lowest edge of my life. I knew my body was physically damaged—actually felt I might die. Then I remember the throwing up, the whole thing.

I see my wife and what my drinking did to her. And my son—what it is still doing to him.

I have a deep moral responsibility for the damage I've done. I've done wrong, morally wrong. The car wreck wouldn't have happened if I hadn't been drinking heavily. I would have reacted differently. My car was a mess. It's sickening. I hit a car head on. Two ladies were in the hospital because of me. I feel like a dog. . . . This could lead to my killing someone or myself being killed. . . . Remember the people after the wreck, looking in the window. There's blood down the front of me, shattered glass all around.

Remember waking up in jail—the blackouts—worrying about whether you'd hit anybody with your car. Your two little boys seeing you drunk.

How nice it would be to have a glass of wine. No!—You'll lose your business—you'll have no chance at this woman—you'll be sick—you'll be out gambling all night. She'll hate you. You'll lose all your money. Everything you're striving for will be down the tubes.

The hangover will be awful. What will I do or what will I say? Will I go into a blackout? Will it be a freaky one? Or will I be a pussycat?

I can see the bottle sitting there and the glass. I remember the effect it had on me and how it tasted. But then I remember the times I was physically sick—throwing up—and the lying I had to do.

Everytime I get the urge to drink, I immediately think of being sick, vomiting, shakes, being miserable.

It would be nice to have a scotch, and I can just about taste it. But I can see myself four scotches later, sloppy, personality changing, not having a pleasant scotch but being drunk.

I'm thinking of this woman I took to the detox center. She stunk, looked like shit. Her skin was horrible—her hair was horrible—she was bloated—broken veins. I don't want to be like that anymore. Looking sad—smelling bad—and sounding ridiculous.

You are going for the blackouts. You're going to get big and fat again. Everybody is going to hate you. You're going to be a failure.

If I'm going to die, I'm going to die trying to quit. It's really killing me and I can see where I've been and what a wretched life I've lived and the thoughts of it are unbearable.

Drinking is poison for me.

If I drink, I'm going to let all those people down. All those people who believed I could stay sober. I'm going to have to go back and face them and tell them I drank, that I didn't make it.

You don't want to drink that. If you do, you are going to be right back in the position you were before. No job. No apartment. Maybe not even a family. I know what can happen and what will happen if I start drinking again. I've been through it so many times. The jobs that have been lost, et cetera. I'm an alcoholic. I can't drink—social or otherwise—not even the first one. I know what I stand to lose if I start drinking.

I don't want to die. I see a real image in my head. A real big, black nothing and it's almost like falling through it. That's what I think of death being like and it pretty well scares the hell out of me. I won't drink tonight.

Fear stays with me—I see people at the VA Hospital—see people in d.t.'s—a sixty-pound man with his mother crying and watching—I can still see those people and feel the fear. I thought I was going to die. Doctor told me I was a short-timer, had cirrhosis. Went to VA. They showed me what could happen—made a believer out of me. Drinking is just like putting a bullet to my head.

With the *benefits-of-sobriety script,* individuals provide themselves with inspirational arguments about the advantages of remaining sober. It is often not enough for individuals to deplore the evils of drink as a means of maintaining abstinence; they need to emphasize the benefits of sobriety as well. This is particularly helpful when they are under considerable stress or in situations conducive to slips. The benefits, naturally, are numerous, covering almost every facet of their lives— improved health, better marital relationships, financial rewards, and a sense of personal freedom.

Here are samples of this type of inner speech.

I can save all that money I throw away. We can go places—take a trip or a vacation or something. I don't have any money problems when I'm sober. I don't have money to do big fancy things and live fancy or anything like that, but I've never been used to having a lot . . . I can't drink—I've got too much responsibility here. I live in the old section, and nearly all the people are getting too old and they don't drive. They look to somebody like me that's got a car to take them to the doctor or somewhere else when they need to go. And sure as hell they don't want to do that with a drunk. . . . You take old women—they blow a light bulb or furnace goes out on them and they come to me for that. I used to wash their windows and such for them. When I'm drinking, I don't have nobody to call on me like that. I'm a good person when I'm sober and I'm pretty wicked when I drink. . . . I've got lots of things to keep on the wagon for if I just listen to the reasons and all.

I remember looking off into the distances and the hills in the background were gray and distant. While I was watching, it was real dark and cloudy, and through the dark cloud there came a shaft of sunlight lighting up this dark place. But it didn't just stay in one place—it moved. The first place was in darkness—this was not. But every time it lit up one place, there was another place in darkness. . . . When I'm in the bad place now [wanting to drink], I can put myself in that dark place I saw, near the mountainside, and I can feel how cold and dark and alone it is. And I can close my eyes and know that just a little bit ahead of me is warmth and sunshine, and it's almost the same kind of warmth and sunshine I felt with alcohol. That's the image that I believe is waiting for me if I just get through this dark place and keep moving on.

There is not a day that goes by that I am not thankful and amazed at how wonderful the whole pattern of things has become—family, financial success, peace of mind, feeling comfortable with myself. Things are so much better now than they were then. I have everything to lose and nothing to gain. [That's what I keep telling myself.]

There is nothing that outweighs my sobriety now. It's everything. If I lose that, I've lost everything. If I lose my sobriety, I've lost me again.

[During a tense and stressful time] Damn it! I'm going to drink. I know that's the only thing to do to relieve this feeling. . . . No way! I've had too much success to go drink . . . Look what I've got. I've got freedom. It's not worth losing all I've got now—the freedom, the joy, the happiness, knowing that I'm an O.K. person. It's not worth it. I have so much invested now. I don't want it. I am free. I have been set free. I've found so much more joy—a fulfillment, a liking myself, sort of feeling in tune with the universe, harmony with the world. And I have it in the natural state. That's what I don't want to lose. No alcoholic beverage gave me what I've finally obtained to this point in my life. I feel a completeness—and love of God and humanity.

I work my butt off for seven years, seven days a week. I have—by society's standards—achieved something. I make a lot by most people's standards. I have a very good reputation in my industry which is nationwide. I've gotten established mainly in the last six years. I would hate to lose my job or money because of the fact that I don't have what it takes to hang in there.

I'm a damn fool to take a drink, so I'm not going to take the chance. For the simple reason that I've got too much at stake. I've got a new home now, I've got a wife and family—three partners and a going concern—and one drink stands between me and all that.

I feel good inside. I feel good about my spiritual relationship with God. I feel good about myself as an individual—that I have capabilities—that I have limitations. It doesn't bother me to tell anyone that I don't know. Well, I don't know. I'm grateful for my sobriety. I appreciate being sober. And I like the good life. I like this kind of life I'm living.

I've gained strength and I feel like an altogether better person—morally, spiritually, and in every other way. I feel good about myself now. I respect myself more. I realize now that I do have some potential and I want to work with that as much as I can. I also want to reach out to help other alcoholics.

I love life now. I'm not worried about tomorrow. I don't have fears in the morning when I wake up.

I don't have to break into my little song-and-dance routine—put on a show for anybody. I don't have to be phony. I don't have to tell lies. I can be what I want to be and what I am—can act the way I am. If people don't like me the way I am, I can stand it either way. If they like me—fine. If they don't like me—fine. I can handle it either way.

I see two images. A flash of what I used to be—a nervous wreck, physically sick. Then a peacefulness comes over me—a peace of mind. Those are the words I use most frequently. I just keep repeating—"I have peace of mind now that I never had while I was drinking."

I know I'm not going to wake up tomorrow morning feeling like hell. I remember the situation I was in at one time. And I know the situation I'm in now. I have a lot of pride.

[I give myself a pat on the back everyday] It's another day and I'm not doing it, and that's great. I feel good. I feel self-confident.

I get along with my wife much better when I'm sober. She can't pick on me as much because I haven't done anything. I don't waste my money either. When I drink steadily, it does mount up—well over $100 a month. The stuff isn't cheap.

I'm a good person. I've got a hell of a good family—a beautiful life. I don't need it.

With the *rationality script,* individuals bolster their resolve to remain sober by mentally telling themselves several, and sometimes, hundreds of times every day that any decision to drink is entirely theirs, that they decided to quit because it was the only sensible thing to do, and that they do have the emotional wherewithal to resist temptation. Opting for sobriety represents a voluntary choice, a rational decision. There is no compelling reason to drink. It's stupid. If they truly don't want to drink, then they really don't have to, no matter what they think or feel. It's a matter of willpower, a statement of choice. And a case of not fooling themselves anymore.

The following samples are representative of this script.

> I have to make up my mind to want to help myself first before anybody can do anything for me, and that's the truth.
>
> I'm not in that much pain. This really isn't going to help.
>
> I'm stupid to try to pull something like this. I know it won't work. I can't possibly hurt anybody but myself if I go on with this game of "I'll show you."
>
> You know, this is a lot of melodrama. I'm not really hurting all that far down.
>
> God gave me another chance with my health. I'm going to take what I have inside me and put it to use. Stop wasting my life. Quit being a nobody. It's a good feeling to be recognized by people [that you have something to offer]. I know I've got it. I've got to make it. I will make it. I want to make it.
>
> I can stay sober if I want to. It's my decision. A man has a responsibility, and he's got to face it. A man has some kind of control over everything.
>
> Most men do have control over their destinies. I hope God will help me out of this situation. I'm going to do the best that I can and I hope He will too. . . . I hope God will help me out, whoever God might be. But if He won't, I've got to do it myself.
>
> I just don't want to take any chances. My mind is set. I'm determined. I don't dare turn back now. There's no way I will turn back now. I just don't want to take that chance. There are too many things I've learned to like about myself. I will find that rainbow someday if I don't screw up too much. And that is too great a risk for me to take.
>
> Well, you told yourself you were going to do it and you've done it. I've accomplished something. I had a problem and I've done it. It's like climbing a mountain. I'm struggling, maybe slipping backwards, forwards, and all that, but I'm making it. I'm on the top and it feels good.

Though reflecting a basic tenet of Alcoholics Anonymous, the *avoid-the-first-drink script*, in some form or another, tends to be used by most recovering alcoholics, whether members of AA or not. For many individuals, the constant reminder that they are indeed alcoholics and cannot control their drinking helps them to check or counter any thoughts about the pleasures of drink or any excuses to test their control. Social drinking or controlled drinking is not an option; total abstinence from alcohol is the only solution. By not taking that first drink, alcoholics, by definition, need not fear relapsing.

This kind of self-talk can take several forms.

> If I don't drink today, tomorrow will take care of itself. Tomorrow will take care of itself as long as I don't drink today.

> I am not going to drink. I made that one commitment, and regardless of what happens I won't.

> I'll never drink again. But I'm going to do it one day at a time. I'm not going to have one today.

> I won't do it. Every morning I say I'm not going to drink any.

> I'm not going to drink. I'm not going to drink. [Over and over] I know if I take it, I'll be on a binge and I don't want to go on a binge. So, I will find myself something to do.

> I'm really an alcoholic and I can't handle it. One drink is all it takes.

> It would be pleasant if I could have a drink, but I know that I won't stop there. And really, things have been so good I don't see how alcohol can enhance my life at all now.

> I'd like a drink. I can take one and stop. On second thought, I know I can't. If I walk away from here, I'll feel much better afterward for not doing it.

> Sure, I'm craving a drink. But one drink will be my undoing. All this that I have built up for two years for myself will be gone down the drain because I'll be flattened again and my chances of ever coming back again will be nil. No matter how bad I want to drink, I know that I can't have any. I know that I'll be right back in the gutter again. I'll be in the snake pit, and I can't handle that trip anymore. I don't want to. I've been sober long enough now to realize how horrible it was and how horrible it is.

> Remember how bad everything has been and how difficult it has been to ever get straight in the first place and how much I have to lose if I

even took one drink. I'm alcoholic and I know I'll never stop at one. I can't get too self-righteous about my position [as an alcohol counselor] because I'm not that far removed from them. Just one drink.

I'm an alcoholic and I know that I have an addictive personality. One drink will not be enough. I'm an alcoholic and I know it. I know what's going to happen.

Don't take that drink, don't take it tonight. Don't pick it up because you know what is going to happen. You are not going to be able to stop. You are going to be sicker than a dog. You are going to miss the appointment in the morning. You really want this job, and if you go in there and mess it up it is going to hurt.

Well, I could take a drink—but the truth is I couldn't possibly control it. I'm going for relief, and relief is good, but I don't think it is worth taking the chance.

I am unable to stop. I'm unable to control alcohol. There is nothing pleasant associated with drinking left in my life.

[After a rough day at work—I'm really tired, sitting in my favorite chair, reading the paper] How nice it would be to go to the freezer and get a cold beer . . . that would be nice just to sit there and enjoy one cold beer. . . . You son-of-a-bitch! Who are you trying to kid? You can't handle one beer. It's time for a change of scenery or something to get that thought out of my mind. If I don't I'll probably get that beer.

I'm not drinking anymore. I know full well what drinking means to me. I know what I do when I start. I know that I can't just have one drink because I've tried that over and over and it doesn't work. I've made a commitment—a pact with myself and I don't want to break it. No matter how strongly I want one of those ice-cold Heinekens. I'm not about to destroy what I have set out to do with one beer.

I have no choice. I have a problem. I can't handle alcohol. It makes an ass out of me. I'm not ashamed. I'm not proud that I'm an alcoholic. But I'm not ashamed. I can't handle it.

If I put six in that refrigerator and I drink the first one, they will all be gone before I quit.

I can never be a social drinker. I'm intelligent. So I'm not going to be any kind of drinker. Just completely stay away from alcohol.

I see myself whirling out of control. Really sick. Just whirling. Space. Nothing. Lost.

Just one drink of alcohol, and I'll be in the gutter again.

One drink would not be just a drink. It would be a drunk.

With a *prayer script*, individuals appeal to a Higher Power to help them think straight. Instead of motivating themselves with inspirational self-talk, individuals gain solace from a personal relationship with God, a sense that they are being looked after from above. The frequency of prayer varies tremendously among individuals, some resorting to prayer only when they feel particularly vulnerable, others praying automatically hundreds of times a day as a talisman against evil. And then there are some who claim to be in a constant prayerful attitude even though they are not consciously praying.

Here are some examples.

I just keep saying the Serenity Prayer over and over each day—the part that says "courage to change the things that I can and the wisdom to know the difference."

I've made up my mind. I really don't want no more booze. I wonder about "Higher Power"—turning things over to Him and not fighting it. I'm turning my wheel of life over to that Higher Power that I choose to call God. "O.K. God. I've tried it time and time again and I haven't handled it. I know you can. You know everything. You are an all-knowing God. You know how I feel." I'll keep in contact with my Higher Power as I understand Him and for wisdom.

Dear God, please don't let me get crazy and continue drinking.

[Constant prayerful attitude] God, don't let me. Thank God for keeping me sober today—or the past hour—or five minutes—or walking past the bar. Thank You for not letting me go in there. Take away the desire to drink. Help me overcome this.

God, you've got to help me now. I'm coming as close as I've come to slipping again. You've got to help me. I still remember that last drunk and how far I've come these last five months without alcohol. I don't want to give it up. Damn, I'm going to make it. I'm not going to get me a drink. I've won a victory!

The wine on the grocery shelf said, "Man, you want me—you'd better buy me. You can really feel good on me. I prayed "God, I'm really miserable. You've got to get me through this."

These are the major kinds of thought processes that seem to be employed by a large segment of recovering alcoholics.[6] But this type of mental programming or self-indoctrination does not always guarantee a problem-free existence. Under stress or duress, individuals have the tendency, especially during the early stages of recovery, to forget their new scripts and revert to old, harmful ways of thought. But if they weather out these times, their task becomes easier and easier. As in acquiring any new skill or language, mastery of these sobriety scripts is dependent upon endless repetition and rehearsal. The more these lines of thought are practiced, the more automatically they come to mind. They not only become more readily available during times of stress, but eventually begin to dominate the entire mental outlook of individuals, crowding out all opposing thoughts. In time, thought patterns that once seemed artificial and alien come to be experienced as habitual and natural, representing a stable pattern of attitudes, assumptions, values, beliefs, and expectations that are associated with sobriety.[7]

But that is not all that is necessary for sobriety. How individuals think must be supported by how they behave. Thought and action constitute a mutually reinforcing system. Sobriety is more than a state of mind. It is a way of life.

9

The Sober Mind

In his imaginative novel, *The Beekeepers*, Peter Redgrove[1] describes the situation of two friends who decide to give up drink after one in d.t.'s heard the doctor say, "Think about bees." The friends take different routes toward abstinence, one cutting alcohol out on his own, the other joining a society called the Institute, which regards the bee as a primitive representation of the soul and is dedicated to substituting trance states for drinking. Members of the Institute spend much of their time caring for the hives or thinking about the bees. The products of the hive—honey, royal jelly, and propolis—are revered as special foods. Tables are polished with beeswax. Recordings of hive activity are used for meditation. And, like a communal hive, members have sex with one another.

I don't know if this description of the Institute is a satire on certain therapeutic agencies, but regardless of its intent, the author identifies a number of essential elements and stages for most successful rehabilitation programs. There is the initial crisis point, in this instance d.t.'s. Then there is the revelatory experience, the quasihallucinatory admonition to "think about bees," which provides the initial incentive for abstinence. Then there is the joining of and total commitment to an organization dedicated to the care, study, and veneration of the bees, a barely disguised symbol of God. Within this society, members consume and use the products of the bees—sober, diligent, busy, and communal creatures—for their daily activities and employ recordings of the hive—comparable to sayings, dogmas, and "buzzwords"—for meditation. If not taken literally, the practice of communal sex can symbolize the likemindedness and closeness among members of this type of organization. At least, that is how I interpret the author's message.

But for our purposes, this fictional account does not go far enough in portraying the process of recovery. What is left unsaid is what happens to recovering alcoholics—whether members of such imaginary organizations or not—once they decide to enter the real world again, but this time without benefit of alcohol. What is the attraction of their new way of life?

What individuals must recognize is that sobriety is like the mythical tree that can bear many different kinds of fruit. Its yield depends largely upon the life circumstances, opportunities, talents, and potential of each individual. But regardless of their situations, sobriety offers all individuals the prospect of a healthier, richer, and more meaningful life than they could ever have had while drinking.[2] However—and this point needs to be made clear—*it does not guarantee it*. Sobriety is not equivalent to a successful life or a smooth, untroubled existence. There is no assurance that individuals will never experience disease, pain, frustration, suffering, rejections, failures, setbacks, or disappointments. What it does promise, though, is that if adversity strikes, individuals will deal with it in a clearheaded and responsible way that allows them to preserve their most precious possession—their self-respect and dignity.

A surgeon, who had been sober for over seven years, says, "There is not a day that goes by that I am not thankful and amazed at how wonderful the whole pattern of things has become . . . family, financial success . . . peace of mind . . . just being able to feel comfortable with yourself. Things are so much better now than when I was drinking."

Another individual who had closed himself off from experiencing positive emotions during his years of drinking, claims that now he allows himself to become sentimental and is "softly pleased" by the affections of his ten-year-old dog, who will just sit and lean his head against his leg.

A thirty-two-year-old newscaster, who had been abstinent for over one year, says, "I've become a lot more active now. Always raring to go, to do something, anything. A baseball game. Volleyball. I'm out there all day long. And eating. I'm loving eating again. I'm one of those fortunate people that can eat and eat and not gain a pound. And as far as stress . . . monetary stress . . . I don't think anyone is without monetary stress, but the stress is much less now. I don't put off paying something. Now, without drinking, the money is there. And I see people all around me get mad at the silliest thing. . . . They fly off the handle and get very upset. And I wish I could tell them . . . 'If you quit drinking, it won't happen. You'd be surprised at the change.' Then I recognize that I didn't want to hear that. So they probably wouldn't want to hear it from a reformed drinker."

And a forty-three-old wallpaper hanger, after eighteen months of sobriety, says, "I feel good about the little things. I feel good because there's sunshine. I feel good because the flowers bloom, and the trees. I feel good because I drive a nice car . . . wear nice clothes. I work very hard on things. I want to be nice to everybody. I want to talk to everybody. I want to meet people now. I want to make sure I remember their names. And I find myself doing things that surprise me. There was a fellow I owed money for some time. I forgot about it. All of a sudden, a couple of weeks ago, I find myself writing a letter to that fellow, saying I was sorry but I want to make it up. So I mailed the letter . . . and I feel good about it."

The testimonials for sobriety are too numerous to recount.[3] But it is one matter to recognize the importance of sobriety and another matter to know how to achieve it. Yet, from what has been described before, the formula for sobriety may not be as complex as it seems.

There are a number of distinct, identifiable ingredients that seem essential for success.[4] At this point, a recapitulation and elaboration of the process of recovery are in order.

As should be obvious by now, the road to sobriety starts with a commitment to see the journey through regardless of the obstacles encountered along the way. Usually, the groundwork for this commitment has been laid over many previous weeks, months, or years—during periods of self-disgust, humiliation, or shame—when individuals kept putting off the inevitable decision until some undetermined, later time. Except for those fortunate individuals, known as "high bottom drunks," who are able to foresee the dangerous ramifications of their drinking long before any real disaster strikes, or those who have been brought to awareness by the forceful intervention of their employers or families, most alcoholics who eventually quit drinking are likely to do so only after a suitable "ripening of motives" that mostly occurs at personal low points or bottoms, moments when they come face to face with the reality of their predicaments, when their defenses have crumbled and they no longer find credible the fictions they have spun to justify what they have done. It is at these decisive moments, usually instigated by a crisis of conscience, that they experience the critical turning point in their lives. They no longer have the luxury of postponing the crucial decision as they had done so many times before. The moment to act becomes now.

Without this deep and abiding will to act, recovery is unlikely. This will represents something more than being aroused for the task. It has a directional component as well. At first, the entire focus is on resisting the urges and temptations to drink. Later, it involves the pursuit of a sober lifestyle and all it entails.

The motivation to quit drinking does not always occur at the moment individuals opt for sobriety, for it is possible that the will to recover may grow slowly in intensity until with the passage of time, it eventually hardens into a firm, implacable resolve. It also is possible that individuals, initially relying on outside assistance, may continue to extract motivational strength from what they perceive to be "all-powerful" helping agencies, like AA, the church, alcohol treatment programs, and even God, until some of this strength can be incorporated within. It is not sufficient, though, for individuals to rely entirely on outside direction for motivational sustenance, participating perfunctorily in certain therapeutic functions or parroting back certain views, as happens in many treatment programs. Helping agencies may not always be there when needed. At some point, either early in their abstinence or later, this will for recovery has to come from within—something they can fall back on when the chips are down—and not be just an ongoing infusion of motivation from elsewhere.

In a way, it seems almost trite to state that alcoholics need to be highly motivated in order to recover. But what is left unstated is the exact nature of this motivation and just how intense it must be. If alcoholics wish to remain abstinent, and particularly so the first several months after quitting drink, they must want to do so more than anything else in the world. Halfhearted commitments don't work. Sobriety must come first in their lives, at times ruthlessly so, even if it means changing important interests, habits, and relationships that have been established over the years.

As most relapsing alcoholics will agree, generating the motivation to quit drinking is a lot easier than sustaining the motivation to stay dry. That is why they can "successfully" quit drinking numerous times but, sooner or later, never seem able to avoid returning to drink. Staying "up" all the time, maintaining their resolve, can be emotionally exhausting. Letting down their guard after a period of time is perfectly natural. What distinguishes successfully recovered alcoholics from all their fallen brethren is not only a firm, unwavering commitment to sobriety but a means for constantly fueling it. The alcoholics who manage to attain sobriety are able to recharge their motivational batteries before they completely run down.

How do they do this? One way is by constantly remembering or regularly reminding themselves about all the bad things that happened to them and their loved ones as a result of drinking. This disgust toward themselves for past transgressions tends to generate a low-grade anger toward anything associated with drinking, a feeling that provides much of the motivational fuel for their commitment. Another way is by constantly complimenting themselves, mentally giving themselves pats on the back, for all their accomplishments—not just for resisting alcohol but also for doing things they can be proud of at home, work, or in the community.

During the early phase of recovery, individuals need to be particularly wary of a false sense of confidence, the feeling that they have their drinking problem licked and that their course in life now seems clear. Still basking in the afterglow of a spiritual experience or floating on cloud nine after a month or two of a relatively trouble-free abstinence does not constitute a suitable basis for successful sobriety. All too often individuals come crashing to earth once they encounter a serious snag. The euphoria understandably experienced during this temporary reprieve from addiction does not provide them with the mental tools necessary for coping with all the tricky and sticky situations that are likely to come their way. Aside from an ongoing resolve for sobriety, what is also needed during this interim period of fragile abstinence is a protective cognitive shield—a mental "germ-free bubble" of sorts—to ward off the countless dangerous thoughts and temptations that can be expected to beset them over the weeks and months ahead.

The usual mental techniques that offer some measure of protection have already been described. Despite its drawbacks, distraction in its many forms is the commonest technique, and work is one of the best forms of distraction. By constantly keeping busy, active, or preoccupied with mentally demanding tasks, individuals can temporarily divert their attention from the unwanted thoughts about alcohol to more constructive pursuits.

The mental repetition of inspirational buzzwords, phrases, or slogans represents another useful technique. These catchy mental statements not only displace the urges and cravings from consciousness, but also serve to "psych" individuals up. Constant prayer is another effective method for drowning out or suppressing errant thoughts. Other potentially effective techniques include substituting negative for positive images, thoughts, or feelings about drinking; thought-stopping to derail unwanted thoughts; and thought-ignoring to extinguish the power of the intrusive ideas. All of these methods serve as types of mental charms, talismans, or amulets to ward off the temptations of drink.

As their urges to drink diminish over time both in intensity and frequency, individuals usually begin to feel more secure in their ability to cope. Now more confident in surviving their ordeal, they can begin to lay the foundations for a more stable world view, one based on the categorical imperative of sobriety. This requires, so to speak, the use of different "mental filters," ones which allow them to screen out all the appeal of a drinking lifestyle and to see themselves and the world from an entirely different perspective—one in which clearheadedness is valued more than intoxication.

Through this period, individuals need to shield themselves against temptation. One of the best ways of doing this is by avoidance. This involves deliberately trying to "put time and distance" between themselves and booze. The farther out of reach and the longer it takes to get alcohol, the more time they have to think about what they are doing. When avoidance of alcohol or former drinking companions is not feasible or practical—for example, when individuals need to attend certain social functions or business luncheons—then they need to prepare themselves beforehand, mentally keeping their guards up, and remaining ever wary of the kinds of incredible "accidents" that tend to befall unsuspecting but susceptible alcoholics, like "inadvertently" picking up someone else's drink. As with defensive driving, anticipation is usually the best way to prevent an accident.

Aside from avoiding risky situations, alcoholics also need to beware of certain vulnerable states of mind. When individuals are feeling sorry for themselves, when they are nursing grievances and resentments, when they are feeling lonely or bored, or when they are engaged in a variety of characteristic thought processes that provide a fertile soil for urges and cravings, they must learn to program themselves with alternative and more constructive ways of thought.

The importance of a positive frame of mind cannot be overemphasized. This is stated well in the old song, which goes,

"You've got to accentuate the positive,
Eliminate the negative,
Latch onto the affirmative,
Don't mess with 'Mister In-between.' "

Though simplistic, these words are particularly apt for alcoholics—or perhaps anyone for that matter. Negative thoughts erode initiative and motivation. Equivocation, indecision, and doubt—the "Mister In-be-tweens"—undermine self-confidence and stimulate the search for certitude and meaning through artificial means or outside sources. Only an optimistic and positive outlook provides individuals with the best protection against adversity and stress.

For some strange reason, many alcoholics have gotten the notion that life without alcohol is a real drag. One alcoholic put this well when he told me "I'd rather be a free drunk than a sober robot." This is nonsense. Partying and the alcoholic lifestyle, especially early on, may be associated with glamour and excitement, but that does not mean, by exclusion, that sobriety has to be equated with solemnity, boredom, and gloom. Just as alcoholism represents a way of life involving more than drinking, sobriety represents a way of life involving more than abstinence. Abstinence is a means, not an end. It is a means toward a more fulfilling and rewarding life. [5]

What needs to be emphasized is the importance of involvement, of participating fully in life. This is the very antithesis of what it was like for individuals when they resorted to alcohol to shield them against certain natural experiences and to alter reality. Theirs was a distorted and artificial approach to life. Instead of being drawn to intoxication, individuals need to get involved in and appreciate the variety of opportunities and experiences that life has to offer. And they need to do so while clear-headed and sober.

Social involvement represents an important component for any positive program of action. Aside from possible membership in organizations like AA, recovering alcoholics can meet most of their affiliative and social needs through participation in temperance groups, church socials, Bible classes, volunteer services, neighborhood associations, or community service organization. Activities like these, along with more intimate encounters with family and friends, should go a long way toward eliminating their long-standing sense of alienation from others and allow them to feel an integral part of a broader social network. Individuals may also find it helpful to get involved in a number of "good causes," activities allowing them to feel they are making a positive contribution to society and thus help them feel better about themselves.

But this is only a part of the solution. Individuals also need to have fun. For many individuals, recreational activities offer significant sources of pleasure, especially if they are embraced in an almost addictive manner. If these healthy habits or "positive addictions," as they have been aptly called,[6] are to have a major impact on individuals, they must eventually come to be regarded both as enjoyable to do and unpleasant to avoid. This is most likely to happen when the activities are noncompetitive, can be done easily without great mental effort, do not depend on the presence or help of others, are believed to hold some physical, mental, or spiritual value, are capable of leading to self-improvement with persistent effort, and can be performed without criticizing oneself. The latter is important in that, if individuals become discouraged with themselves, the activity will lose its pleasure, and without its being pleasurable, it will not be "addicting." Activities such as jogging, weight lifting, swimming, aerobic exercises, meditation, or archery seem to fill the bill admirably.

Not all activities need have an addicting or compulsive quality for them to be embraced enthusiastically by individuals. Hobbies and "substitute indulgences"[7] may provide a good deal of enjoyment and intellectual stimulation. Light reading, movies, drawing, massages, saunas, chess, computer games, stamp collecting, rock polishing, museum going, motorcycling, spelunking, canoeing, camping, hunting, learning about nutrition, gourmet cooking, gardening, woodwork, knitting, crocheting, quiltmaking, model building, photography, and a host of other activities[8] can provide individuals with the necessary pleasure and relaxation to restore or maintain proper emotional balance.

All of the techniques and activities so far described represent the building blocks for a successful recovery. But what eventually cements them together is the commitment to an emotionally, intellectually, and spiritually fulfilling life. This is the hallmark of true sobriety. Countless tomes could be written about how this is to be achieved, but none, in my opinion, can possibly offer a better prescription than that contained in The Serenity Prayer—even for those who have trouble accepting the notion of a Higher Power.

(God) grant me the serenity to accept the things I cannot change,
the courage to change the things I can,
and the wisdom to know the difference.

The simple sentiments expressed in these relatively few words have profound implications for human behavior. As I interpret these words, they mean facing what needs to be faced. They mean avoiding boredom. They mean dealing with anger and resentments. They mean being able to tolerate frustration. They mean avoiding rationalizations and self-deceptions. They mean accepting personal limitations. They mean risking disapproval. They mean being empathetic and loving. They mean resolving conflicts when they arise. They mean making enlightened decisions. They mean taking responsibility for personal behavior. They mean coming to peace with oneself. And above all, they mean dealing with what life has to offer and getting involved in the process of living.

With this orientation to life, intoxication is unnecessary.

Notes

Preface

1. In my opinion, it is helpful to conceptualize "alcoholism" as embracing all the features of a drug-dependence syndrome. The concept of the drug dependence syndrome was developed by the World Health Organization's program on nomenclature and classification of alcohol and drug-related problems. The core syndrome contains cognitive, behavioral, and physiological features which are presumed to be related to a common psychobiological process, which vary in severity, and which are influenced by cultural and personality features. (For further details see Edwards et al., "Alcohol-related problems"; Edwards et al., "Nomenclature and classification"; Edwards and Gross, "Alcohol dependence.")

The key features of this drug-dependence syndrome are as follows:

Narrowing of the repertoire of alcohol consumption. With the exception of binge drinkers, there is a tendency for the drinking pattern to become more stereotyped over time, with little variation in beverage choice, drinking times, and daily alcohol consumption.

"Drink-centeredness" or salience of drinking behavior. Alcohol con-

sumption tends to be given higher priority than other important activities despite its negative physical and social consequences.

Tolerance.   More and more alcohol is required over time to produce the same subjective and behavioral effects.

Physical withdrawal symptoms.   As alcohol use increases, sudden drops in alcohol intake or blood alcohol level tend to be associated with physical withdrawal symptoms such as shakiness, agitation, insomnia, and confusion.

Use of alcohol to avoid withdrawal symptoms.   Alcohol tends to be used to avoid the unpleasant symptoms associated with alcohol withdrawal.

The compulsion to use alcohol.   Individuals tend to experience a growing compulsion, desire, or craving for alcohol, especially during attempts to curb its use, as well as impaired ability to reduce the amount consumed.

Readdiction potential.   As individuals experience a growing lack of control over the frequency and amount of alcohol intake or experience its negative consequencies, they may desire to quit drinking on a temporary or even permanent basis. However, following a variable period of abstinence, individuals have a tendency to revert rapidly to old drinking patterns after beginning to drink again.

(See R. Straus, "Types of alcohol dependence" and "Alcohol and alcohol problems" for an interesting perspective on the issue of alcohol dependence.)

2. Most of what I have to say applies equally well to both sexes. Unfortunately, because of the limitations of the English language, which make the use of "he and she" or "s/he" stylistically awkward every time a reference to an alcoholic is made, I have decided to use the pronoun "he" in its generic sense to refer to both sexes. Fortunately, this problem doesn't come up that often.

Chapter 1

1. There are alternative and perhaps better ways of organizing notions about alcoholism than that which I propose. Philip Brickman and colleagues ("Models of helping and coping") describe four different models of alcoholism based on the concept of responsibility. With the *moral model*, the alcoholic is responsible both for his problem and its cure. With the *medical model*, the alcoholic

is responsible for neither since alcoholism is regarded as a disease. With the *compensatory model*, the alcoholic, who is presumed to be handicapped or disadvantaged in some way, is not held responsible for his problem but he is for its cure. And with the *enlightenment model*, the alcoholic is responsible for his problem but, because of the intervention of a Higher Power, AA, or other outside, authoritative agency, not for its cure. In contrast, Siegler and Osmond *(Models of Madness)* propose eight models of alcoholism—the impaired model, the "dry" moral model, the "wet" moral model, the AA model, the psychoanalytic model, the family interaction model, the "old" medical model, and the "new" medical model—each with its own implications for etiology and treatment. Though I have adopted elements from both of the above conceptualizations, I have found my own approach to be more useful since it is somewhat simpler and involves different categories of information and facts.

Chapter 2

1. London, *John Barleycorn*, p. 209.
2. See account by Elmore Leonard in Wholey, *The Courage to Change*.
3. Marlatt refers to related phenomena as "apparently irrelevant decisions," a particularly apt term (see Marlatt and Gordon, *Relapse Prevention*). In alcoholic parlance, they are referred to as "slippery thinking."
4. This is a direct quote from one of my alcoholic informants.
5. See Marlatt, "Craving for alcohol"; Marlatt, "A cognitive-behavioral model"; Marlatt and George, "Relapse prevention"; and Marlatt and Gordon, *Relapse Prevention*.

According to G. Alan Marlatt, Director of the Addictive Behaviors Research Center at the University of Washington, there is another factor at work that contributes to the process of relapse. This has to do with the general assumption that a single drink, in a lock-step fashion, will inevitably lead to drunkenness ("Drink, drank, drunk," so the AA saying goes), and that drunkenness is equivalent with relapse. This leads to a self-fulfilling prophesy in which everyone, including the alcoholic himself, automatically regards the alcoholic as good as gone once he takes that first drink. Sobriety represents an all-or-none phenomenon. There is no room for error. Since a single drink violates the rule of abstinence and, once taken, cannot be undone, then the alcoholic might just as well do it up right.

This expectation, in which a drink is equated with a drunk and a drunk with relapse, has been referred to as the "abstinence violation effect." In Mar-

latt's scheme, a characteristic sequence of events occurs in the process of relapse. The alcoholic is exposed to a high risk situation but doesn't cope adequately. This leads to a diminished sense of "self-efficacy" which, in turn,
prompts the alcoholic to seek relief from this unpleasant, subjective state by
the consumption of alcohol. Once the individual takes the first drink, the
abstinence violation effect takes over, guaranteeing a full-blown relapse.

There are two components to the abstinence violation effect. The first is
known as "cognitive dissonance," a term coined by Festinger (A *Theory of
Cognitive Dissonance*) to denote an intolerable mental state in which contradictory or conflicting notions or expectations coexist, automatically prompting
the individual to seek resolution. In the case of the alcoholic, the taboo behavior, representing the initial slip, is dissonant with his self-image of being
abstinent. The attitude of "I shouldn't have done it, but I did," can be expected to lead to a state of inner turmoil, conflict, and guilt. Along with these
unsettling feelings, the "personal attribution effect" comes into play. Because
of a natural need to explain why he did what he did and why he now feels as
he does, the alcoholic blames his slip on personal weakness or failure rather
than on the situational or environmental factors that put him in the high-risk
situation. Regarding himself as lacking in willpower, as having virtually no
control over his behavior, the alcoholic resigns himself to the inevitable, and
a full relapse takes place. Then, after a drinking bout of variable duration, the
cycle begins anew with the switch from complete loss of control to absolute
control over drinking, usually initiated by outside pressures or factors, like
compulsory treatment, severe illness, "hitting bottom," legal sanctions, or a
religious experience.

On the basis of these theories, Marlatt and colleagues propose a highly
innovative program for relapse prevention. Assuming that forewarned is
forearmed and that "slips" for most alcoholics are inevitable, they maintain
that alcoholics should be taught how to cope with a slip constructively rather
than capitulating entirely to the abstinence violation effect. Drunkenness need
not be an inevitable consequence of drinking. One of the many techniques
employed to implement this notion is mental rehearsal, during which individuals are encourged to picture themselves handling a slip in more effective ways
than they have in the past.

The logical extension of this technique is a "programmed relapse," during
which alcoholics in treatment are expected to consume the first drink under
the supervision of a therapist. During this time, the therapist attempts to facilitate cognitive restructuring and teach coping strategies for avoiding further
alcohol consumption. The first thing the alcoholic is taught is that a slip
represents a mistake rather than a relapse. Rather than believe himself to be
weak or lacking in willpower when he takes that first drink, the alcoholic

should regard his behavior as caused by external circumstances, and hence, potentially controllable. The lapse represents only a specific event in time and place and, as such, can be avoided in the future. Abstinence, after all, is only a moment away. It only involves the decision to stop drinking. All good advice, if the alcoholic follows it.

This ambitious, self-control program combines behavioral skill training, cognitive interventions, and lifestyle change procedures. Aside from learning early-warning signals and engaging in "relapse drills," alcoholics are exposed to relaxation training, stress management techniques, the formulation of decision matrices, and educational sessions aimed at cognitive restructuring. Not all treatment is directed toward dealing with alcohol per se. Global self-control strategies aimed at a balanced lifestyle, which include exercise, relaxation, and healthful dieting, are encouraged. (For a more complete description of this program, see Marlatt and Gordon, *Relapse Prevention*, Chapters 1 and 2.)

Though credit must be given to Marlatt and colleagues for their insightful and innovative ideas about relapse, I confess to serious reservations about the basic premises. Granted, the expectation among alcoholics and others that the initial violation of abstinence will likely lead to a complete relapse seems absolutist, aribtrary, and unreasonable, and may prove to be a self-fulfilling prophesy, but the real question is whether it is valid. Is the expectation really an unfounded assumption without scientific basis, or is it the distilled wisdom of countless alcoholics and clinicians who have learned, after hard experience, that "slips" are not only dangerous but usually indicative of relapse? My own clinical experience indicates the latter premise to be true. That does not mean that the abstinence violation effect is not an important factor to be reckoned with or that the first drink need inevitably lead to loss-of-control drinking and prolonged drunkenness. It merely means that the accepted notion of a "slip" is short-sighted and faulty. By definition, a "slip" is more than a slip if it usually leads to an extended period of drinking. The reason that most alcoholics relapse after taking the first drink or two is not that they violate an expectational set and then regard themselves as weak-willed and out of control, as Marlatt might claim, but because at some level of consciousness, whether they admit it to themselves or not, they already have committed themselves to get drunk again, to re-experiencing the golden glow of intoxication. It may be on the first few occasions when they are trying to modulate their drinking, playing games with themselves to prove that they are really in control, or later when they say "The hell with it" and finally do what they have wanted to do all along. In other words, the reason that alcoholics have difficulty stopping after the first drink, why a slip is so often indicative of a relapse, is simply because they don't want to stop. Why bother to slip unless you can do it right? Sure, there are exceptions. Rare individuals may recover from a slip or even manage to drink moderately for an extended period of time, but they are engaging in a perilous game, walking along the edge of a precipice, in their effort to prove to themselves and others that they really are in control. One misstep, and they are back where they started.

6. See M. Only, *High*, p. 149.

7. One of my alcoholic patients once told me that the secret dream of every alcoholic is to become a social drinker, to relinquish the pledge of abstinence for the assurance of controlled drinking. Many may take exception to this claim, but judging by the high rate of relapse after treatment, I suspect that my patient may not be too far off. Most alcoholics return to drinking with the expectation that this time they will be able to drink socially. But what an alcoholic means by social drinking is substantially different from what a social drinker means by social drinking. Social drinking, for this particular patient, meant the ability to consume two or three drinks in an evening, like everyone else. Each of these drinks, though, would contain a minimum of four ounces of liquor!

This peculiar notion about social drinking is important since it gets to the heart of the problem with "slips." For most alcoholics, the slip is not an excuse to taste alcohol again but an opportunity to reexperience a "buzz" or "high." That's mainly why alcoholics "can't" stop after one drink. Those authors who argue that many can return to normal drinking (see Heather and Robertson, *Controlled Drinking*) fail to grasp an essential point: it is less frustrating for the preponderance of alcoholics to avoid drinking altogether than to settle for one normal-sized drink, such as a single martini, a glass of wine, or a beer.

The alcoholic's liberal view of social drinking receives support from a most unlikely source. The controversial RAND report (Armor et al., *Alcoholism and Treatment*), which indicates that 12 percent of treated alcoholics were able to return to "normal drinking" after six months and 22 percent after eighteen months, employed most interesting criteria for this conclusion. Normal drinking was defined as less than an average of three ounces of ethanol per day and less than five ounces on drinking days. Translated, these figures mean a total of 21 ounces of absolute alcohol or 42 ounces of 100 proof alcohol or 52 ounces of 80 proof alcohol per week. That's an average of five 1¼- to 1½-ounce martinis or about a six-pack of beer a day, if my calculations are correct, or about eight martinis on special drinking days. If this is "normal" or moderate drinking, then ordinary social drinkers will have to be reclassified as teetotalers!

As to whether the goal of total abstinence is unreasonable and arbitrary, I happen to think not, despite the difficulty alcoholics have in obtaining it. I see little virtue in minimizing the dangers of a slip by offering the illusion of self-control after the first drink when clinical experience, again and again, dictates that the vast majority of alcoholics cannot or do not display this ability. There is wisdom in the AA injunction, "Avoid the first drink." A slip is a serious setback, not just an innocent mistake, mainly because it is more than a slip. It is a reflection of a deep-seated desire to drink again, a rejection of the importance of sobriety.

Chapter 3

1. The concept of a "dry drunk" has been popularized largely by AA. Though this concept has no precise scientific definition or even legitimate clinical status, it nevertheless represents, in my opinion, an important determinant of relapse.

2. In its broadest sense, the notion that certain kinds of thought processes tend to be associated with addictive behavior or other forms of irrational or deviant behavior receives support from the work and writings of various cognitive-behavioral theoreticians (selected references include Meichenbaum, *Cognitive-Behavior Modification*; Ellis, *Reason and Emotion*; Maultsby, *Rational Behavior Therapy*; Beck et al., *Cognitive Therapy of Depression*, Bandura, "Self-efficacy"; and Mahoney, *Cognition and Behavior Modification*).

3. For this chapter only, I have exerted a degree of artistic license in artificially integrating the actual thought processes from numerous alcoholics into separate, reasonably coherent mental scripts which seemed suitable for each of the individuals mentioned.

Chapter 4

1. Bruccoli, *Selected Letters of John O'Hara*, pp. 491–492.

2. See Mello, "Behavioral studies in alcoholism." This argument is a clear example of a logical straw man with little factual basis, since few sophisticated physicians would claim that uncontrolled drinking is an inevitable consequence of craving. See my article (Ludwig and Wikler, "Craving and relapse") for a detailed discussion of this topic.

3. See Mello, "Some aspects of the behavioral pharmacology of alcohol"; Mello and Mendelson, "Operant analysis of drinking patterns"; Mello, McNamee, and Mendelson, "Drinking patterns of chronic alcoholics"; Cohen, Liebson, and Faillace, "The modification of drinking"; Davies, "Normal drinking"; Bailey and Stewart, "Normal drinking by persons"; and Merry, "The 'loss of control' myth."

4. Musgraves and Balsiger, *One More Time*.

5. See Jellinek et al., "The 'craving' for alcohol," p. 63.

6. Isbell, "Craving for alcohol."

7. Related to the concept of symbolic craving is the presumed phenomenon of loss of control. This is commonly·understood as meaning that a drink or two of an alcoholic beverage, taken either wittingly or unwittingly by an alcoholic, will trigger an inordinate desire, a sort of chain reaction, for alcohol, leading to increased consumption and ending with drunkenness or stupor—a process over which the alcoholic has little control. Some qualifications of this rather extreme, but generally accepted, view of loss of control are necessary. There may be an individual threshold, a critical blood alcohol level, "above which *loss of control* begins to set in, a level range varying from person to person, and in the same individual from time to time, depending on a variety of factors, which might be predominantly psychological and/or physiological, psychosomatic or psychosocial in nature." These factors might include (a) state of mind of the alcoholic, (b) motivation for taking the drink, (c) place of drinking and one's company, (d) certain brakes, such as responsibility, work, lack of money, and (e) occupation and environment (See Glatt, "The question of moderate drinking").

In Jellinek's view *(The Disease Concept of Alcoholism)*, craving and loss of control are intimately related. He believes that the obsessive drinking bout due to loss of control becomes partially established through a true physical demand for alcohol (adaptation of cell metabolism to that substance) in the given situation, and that the demand for alcohol gives rise to the idea of "craving." The craving, however, may be only an apparent one since it will not pertain specifically to the ingestion of alcohol, which may be viewed by the alcoholic with disgust, but rather to the expected relief from the painful withdrawal symptoms. "The demand for alcohol seems to be of a two-fold nature. One part reflects the necessity to allay the distressing withdrawal symptoms; i.e., a physical demand; the other part reflects the obsessive belief that ultimately a sufficient amount of alcohol will bring about tension reduction which, before loss of control, was achieved quite easily."

In an attempt to account for loss of control, clinicians have advanced a variety of explanations, including operant and classical conditioning mechanisms, the activation of specific hypothalamic centers eliciting craving, alcohol-induced dissociation of control centers in the brain, and alterations in cellular metabolism that become conditioned by the first drink. At the other extreme, some have claimed that loss of control doesn't exist. For the present, it is sufficient to note that none of these possibilities has been systematically studied or documented.

8. Ludwig and Stark, "Alcohol craving."

9. Hughes, *The Man From Ida Grove*, pp. 114–115.

10. Ludwig and Lyle, "The experimental production of narcotic drug effects."

11. The classic treatise by MacAndrew and Edgerton *(Drunken Comportment)* documents the profound effects of social expectations on the behavior of individuals when they are drinking.

12. See Marlatt and Rohsenow ("The think-drink effect") for details of their study.

13. Ludwig, Wikler, and Stark, "The first drink"; Ludwig et al., "Physiologic and situational determinants."

14. This observation is consistent with that of Meyer and Mirin *(The Heroin Stimulus)*, who claim that the potential availability of an addicting drug like heroin plays an essential role in the experience of narcotic drug hunger.

15. See Ludwig, "The irresistible urge."

16. Dr. Abraham Wikler was a pioneer in the area of drug addiction. Most of what is known about classical conditioning apsects of narcotic use derives directly from his early work. See Wikler, "Recent progress in research"; Wikler, "A psychodynamic study"; Wikler, "A rationale of the diagnosis and treatment of addictions"; Wikler, "On the nature of addiction"; Wikler, "Conditioning factors in opiate addiction"; Wikler and Pescor, "Classical conditioning of morphine-addicted rats"; Wikler, "Interaction of physical dependence and classical and operant conditioning"; Wikler, "Some implications of conditioning theory"; Wikler et al., "Persistent potency of a secondary (conditioned) reinforcer"; and Wikler, "Sources of reinforcement for drug-using behavior."

17. See Wikler, "Dynamics of drug dependence"; O'Brien, "Experimental analysis of conditioning factors"; Grabowski and O'Brien, "Conditioning factors in opiate use"; and Lynch et al., "Pavlovian conditioning of drug reactions."

18. See O'Brien, "Experimental analysis of conditioning factors"; O'Brien et al., "Conditioning of narcotic abstinence symptoms"; and Ternes et al., "Conditioned drug responses to naturalistic stimuli."

19. Eriksen and Götestam, "Conditioned abstinence in alcoholics."

20. The "opponent-process" theory of acquired motivation offers an excellent explanation of how various environmental events may become conditioned to the pleasurable and withdrawal effects associated with repeated alcohol use. (See Solomon, "The opponent-process theory of acquired motivation"; Solomon and Corbit, "An opponent-process theory of acquired motivation"; and Solomon, "An opponent-process theory of acquired motivation: IV.") On the assumption that organisms act to moderate the emotional responses to novel stimuli, the theory proposes two automatic processes, *a* and *b*. The *a* process is triggered by stimuli of all sorts—smoking, parachute jumping, a

reward, or the consumption of alcohol—and leads to an emotional response A. This automatically stimulates a "slave" *b* process that produces an emotional state B that acts to oppose and alter the A state. This keeps the individual protected from extreme emotional reactions to repetitive stimuli.

In the case of early alcohol use, the theory maintains that initial alcohol ingestion should lead to a pleasurable emotional A state which, after some delay, would be opposed by the *b* process. The *b* process becomes particularly strong during a drop in the blood alcohol level, with a resultant B state that is unpleasant. This is pretty much the case with social drinkers. With prolonged, repeated use of alcohol, though, the magnitude of the positive A state presumably gets less and its duration diminishes, while the *b* process increases in strength and length, resulting in a stronger negative affective state. This situation parallels the situation in alcoholics who need to keep drinking just to feel normal.

With respect to the matter of relapse, the opponent-process theory proposes that neutral stimuli can become conditioned to the A and B emotional states through repeated pairings. Stimuli usually present during the early phase of drinking would come to be positive reinforcers since they could elicit pleasurable A states. In time, though, as the aversive B state becomes more prominent, stimuli such as anger, anxiety, or apprehension may be confused by the individual as withdrawal symptoms and elicit craving and alcohol consumption as a negative reinforcement to these effects. This also pertains to any environmental or psychophysiological stimuli that have been associated with withdrawal.

An excellent analysis of the relationship of the opponent-process theory to our work on craving is given by Donovan and Chaney, "Alcohol relapse prevention." See also my chapter, "Why alcoholics drink," for a more comprehensive overview of the various factors associated with relapse.

21. The relationship of craving to degree of physical dependence has also been found by others. Hodgson, Rankin, and Stockwell ("Alcohol dependence and the primary effect") found that craving for alcohol was much higher in severely dependent than moderately dependent alcoholics three hours after a primary dose of alcohol. These investigators also found that severely dependent alcoholics, compared to moderately dependent alcoholics, consumed a fixed dose of alcohol more rapidly and drank more alcohol over a thirty-minute interval (Rankin, Hodgson, and Stockwell, "The behavioral measurement of alcohol dependence"). Moroever, they also had more difficulty in subsequently resisting a drink (Stockwell et al., "Alcohol dependence, beliefs and the primary effect"). In a questionnaire study, Mathew, Claghorn, and Largen ("Craving for alcohol") found that the frequency and intensity of the craving experience decreased with the duration of sobriety—the longer the duration,

the less the craving. The decision to consume alcohol can also be predicted by recent withdrawal symptomatology, as well as an increased heart rate in the presence of alcohol (Kaplan, Meyer, and Stroebel, "Alcohol dependence and responsivity").

22. See Ludwig and Stark, "Alcohol craving." Comparable findings of 85 percent and 58 percent were obtained by Mathew, Claghorn, and Largen ("Craving for alcohol"). It also should be noted that Tokar et al. ("Emotional states and behavioral patterns") reported that alcoholics, in comparison to normal controls, were most likely to drink, smoke, and take pills whenever they felt helpless, depressed, angry, and anxious.

23. See account in Wholey's book, *The Courage to Change*.

24. Ludwig, "Pavlov's 'bells.' "

25. There is evidence to indicate that heavy drinkers find more reasons than light drinkers to justify their drinking. Social reasons, for example, include celebrating special occasions, being sociable, drinking because it's the polite thing to do or because others do it. Escape reasons include relaxation, the relief of tension, cheering up, and forgetting worries. And miscellaneous reasons include liking the taste and improving the appetite. These results have been reported by Cahalan, Cisin, and Crossley in *American Drinking Practices*.

Chapter 5

1. See *Living Sober*.

2. See Victor and Adams, "The effect of alcohol."

3. See Kissin, "The role of physical dependence and brain damage."

4. Some clarification is needed about this statement. Though the actual percentage of alcoholics who recover as a result of treatment is small, the total number who recover has to be regarded as "significant," considering the large numbers of alcoholics involved. A 5 percent recovery rate for several million alcoholics, for example, involves a lot of people.

5. See Smart, "Spontaneous recovery." In another survey, Blane ("Issues in the evaluation of alcoholism treatment") indicates six-month abstinence rates from a wide variety of treatment modalities from 17 to 35 percent and from no or minimal treatment from 13 to 16 percent. In a recent follow-up study of 1,289 alcoholics treated at various medical and psychiatric treatment facilities, Heltzer et al. ("The extent of long-term moderate drinking") found that only 15 percent achieved total abstinence over the prior three years. In addition, only 1.6 percent were able to achieve stable moderate drinking at follow-up, indicating the relative rarity of this outcome after treatment.

6. See Ludwig, Levine, and Stark, *LSD and Alcoholism*. It is of interest to note that the latest RAND report by Polich et al., also revealed that only 7 percent of a much larger sample of alcoholics, an identical percentage to ours, managed to achieve sobriety throughout the entire four-year follow-up period.

7. See Ludwig, Levine, and Stark, *LSD and Alcoholism*.

8. See Keller, "On the loss-of-control phenomenon."

9. See Vaillant, *The Natural History of Alcoholism*.

10. See Lee, *Journey Into Nowhere*.

11. Gerard, Saenger, and Wile, "The abstinent alcoholic."

12. It is interesting to note that recovering alcoholics seldom credit prior treatment for their abstinence (see Ludwig, "On and off the wagon"; Edwards et al., "Alcoholism"; Saunders and Kershaw, "Spontaneous remission"). The initiating causes of abstinence are discussed in Chapter 6.

13. Reports from Raleigh Hills Hospitals indicate impressive twelve-month abstinence rates for their chemical aversion counterconditioning program, ranging from 39 to 63 percent (see Neubuerger et al., "Abstinence following counterconditioning"; Wiens and Menustick, "Aversion therapy"). Unfortunately, like so many other impressive reports, these are not controlled studies. See also Riddell (*I Was an Alcoholic*) for a personal account of exposure to this treatment. A comprehensive overview of alcoholism treatment in the Soviet Union can be found in Babayan and Gonopolsky, *Textbook on Alcoholism*.

14. See Madill et al., "Aversion treatment of alcoholics by succinylcholine."

15. Laverty, "Aversion therapies."

16. This theoretical approach to the treatment of narcotic addiction was first advocated by Wikler (see "Dynamics of drug dependence").

17. See Kwentus and Major, "Disulfiram."

18. See Lee, *Journey Into Nowhere*, p. 112.

19. See Overton, "State dependent learning."

20. Brick's classic statement to his father (Big Daddy) is as follows: "This click that I get in my head that makes me peaceful, I got to drink till I get it. It's just a mechanical thing, something like a . . . switch . . . in my head, turning the hot light off and the cool night on and . . . all of a sudden there's peace!" (see Williams' *Cat on a Hot Tin Roof*, p. 98).

21. See Emrick, "A review of treatment of alcoholism: I"; and Emrick, "A review of treatment of alcoholism: II." This latter review of 384 treatment studies showed that differences in treatment methods did not significantly affect long-term outcome. The average abstinence rates did not differ between treated and untreated alcoholics, but more treated than untreated alcoholics

improved. This higher improvement rate for treated alcoholics is open to many interpretations, one of which is that it represents an artifact of the sampling technique.

Though a variety of behavioral therapies have been advanced for alcoholism, including covert sensitization, contingency contracting, marital skills training, assertiveness training, and a broad spectrum treatment approach, these therapies, while probably as effective as conventional methods, have not produced any significant treatment breakthroughs. See overviews by Litman and Topham, "Outcome studies in alcoholism treatment"; Caddy and Block, "Behavioral treatment methods for alcoholism"; and Baekeland, "Evaluation of treatment methods."

22. See Ludwig, Levine and Stark, *LSD and Alcoholism*.

23. Cooney, Baker, and Pomerleau ("Cue exposure for relapse prevention") describe a clever variant of this approach in which a group of alcoholics was requested to role-play actually turning down a cold beer. This cue exposure technique, designed to deal with a real life situation, generated a wide range of interesting reactions from patients. Unfortunately, this approach has not been evaluated systematically on a long-term basis.

24. See Wilson ("Alcoholics Anonymous—beginnings and growth"). It is difficult to evaluate this claim scientifically, particularly since long-term controlled, follow-up studies are not feasible. Tournier ("Alcoholics Anonymous as treatment and as ideology") also maintains that recovered alcoholics may gravitate toward AA more as a means of sustaining recovery, as a form of aftercare, rather than a specific treatment for initiating sobriety.

25. See Baekeland, "Evaluation of treatment methods"; Tournier, "Alcoholics Anonymous as treatment and as ideology"; Hoffman, Harrison, and Belille, "Alcholics Anonymous after treatment."

26. See Ludwig, Levine, and Stark, *LSD and Alcoholism*.

27. Edwards et al., "Alcoholism."

28. Rudy, "Slipping and sobriety."

29. I have borrowed heavily from Bean's excellent overview ("Alcoholics Anonymous") for much of my commentary about AA, since her views reflect my own. See also Beckman, "An attributional analysis"; Kurtz, *Not-God*; Maxwell, *The A.A. Experience*; and Rudy, *Becoming Alcoholic*, for additional perspectives on AA.

30. The Twleve Steps of AA are as follows (see *Twelve Steps and Twelve Traditions*):

We admitted that we were powerless over alcohol—that our lives had become unmanageable.

Came to believe that a Power greater than ourselves could restore us to sanity.

Made a decision to turn our will and our lives over to the care of God *as we understood Him*.

Made a searching and fearless moral inventory of ourselves.

Admitted to God, to ourselves, and to another human being the exact nature of our wrongs.

Were entirely ready to have God remove all these defects of character.

Humbly asked Him to remove our shortcomings.

Made a list of all persons we had harmed, and became willing to make amends to them all.

Made direct amends to such people whenever possible, except when to do so would injure them or others.

Continued to take personal inventory and when we were wrong promptly admitted it.

Sought through prayer and meditation to improve our conscious contact with God *as we understood Him*, praying only for knowledge of His will for us and the power to carry that out.

Having had a spiritual awakening as a result of these steps, we tried to carry this message to alcoholics, and to practice these principles in all our affairs.

31. Rudy, *Becoming Alcoholic*.

32. It has been suggested that the very issue of AA effectiveness as a treatment is irrelevant since it is not so much of a therapeutic program as a philosophy of recovery. Kurtz ("Why A.A. works") offers an elaborate discussion of the existential underpinnings for this organization based on two fundamental assumptions. One, AA members are not infinite, not absolute, *not God*; acceptance of their personal limitations serves as a beginning for healing and wholeness. Two, AA member are connected with other members in a bond of fellowship.

33. See Bean, "Alcoholics Anonymous."

34. See Goodwin, *The Writer's Vice*.

35. Galanter, "Sociobiology and informal social controls"; Galanter and Buckley, "Evangelical religion"; Simmonds, "Conversion or addiction."

36. Galanter, "The 'relief effect.' "

37. Greil and Rudy, "Social cocoons."

Chapter 6

1. See London, *John Barleycorn*, pp. 277–278.

2. See James, *The Varieties of Religious Experience*, p. 377.

3. This is a quote from a letter by Carl G. Jung to Bill Wilson about Rowland H. and appears in the book by Thomsen, titled *Bill W.*

4. Information about the initiation of recovery is limited and is confounded by inconsistent and ambiguous terminologies among the various authors. One finding agreed upon by almost all investigators is that recovered alcoholics seldom credit prior psychiatric counseling, hospital treatment, or remarkably, even Alcoholics Anonymous, for their initial decision to become sober (see Edwards et al., "Alcoholism"; Gerard, Saenger, and Wile, "The abstinent alcoholic"; Knupfer, "Ex-problem drinkers"; Saunders and Kershaw, "Spontaneous remission"; Cahalan, Cisin, and Crossley, *American Drinking Practices*; McCance and McCance, "Alcoholism in Scotland"). In a retrospective study of 500 alcoholics, Lemere ("What happens to alcoholics") found that 22 percent quit drinking because of a serious physical illness and another 11 percent also became abstinent, two-thirds of whom did so entirely on their own volition and the remainder with the aid of medical treatment, religion, or AA. In my own eighteen-month follow-up study on 176 alcoholics (Ludwig, "On and off the wagon"), only 5 percent of the 94 alcoholics who managed to attain at least one three-month period of abstinence credited prior hospital treatment, and 7 percent credited Alcoholics Anonymous for their decision to "go on the wagon." In a large community sample of alcoholics, Vaillant and Milofsky ("Natural history of male alcoholism") found that self-help was more useful than clinical treatment. In his book, *The Natural History of Alcoholism*, Vaillant concluded that "alcoholics recover not because we treat them but because they heal themselves" (p. 314).

The initiation of sobriety without the aid of formal psychotherapy, hospital treatment, or Alcoholics Anonymous has been referred to as "spontaneous remission." This is a somewhat ambiguous term since it implies that alcoholism, like some other serious medical diseases, can undergo remission for variable lengths of time and for no recognized cause. Knupfer ("Ex-problem drinkers") found a 30 percent recovery rate, three-quarters of which were "spontaneous" in nature. In the most exhaustive literature review on this topic, Smart ("Spontaneous recovery") indicated that cumulative rates for spontaneous recoveries vary from 10 to 42 percent in different studies, with yearly rates ranging from 1 to 33 percent. Alcoholics treated for physical illness as a result of drinking tend to have the highest rates of recovery. Vaillant and Milofsky ("Natural history of male alcoholism") found that 10 to 20 percent of alcoholics never relapse after their first serious request for help, and afterwards 2 to 3 percent will achieve a stable recovery each year. Drew ("Alcoholism as disease") attempted to account for the declining incidence of alcoholism with advancing age on the basis of a maturing-out process and claimed that the data supported the notion of a "self-limiting disease."

Among the various follow-up studies (Edwards et al., "Alcoholism"; Knupfer, "Ex-problem drinkers"; Saunders and Kershaw, "Spontaneous remission";

Smart, "Spontaneous recovery"), the most frequent reasons given for the initiation of remission involved changes in life circumstances, such as marriage, job, or residence, as well as deterioration in physical health. In my study (Ludwig, "On and off the wagon"), the three most common reasons for abstinence were loss of desire for alcohol, fear of the consequences of drinking, and insight into problems. Recently, Tuchfeld ("Spontaneous remission in alcoholics") interviewed fifty-one alcoholics and found that seventeen attributed the initiation of their recoveries to personal illness or accident, six to education about alcoholism, thirteen to religious conversion or experience, nine to direct intervention by immediate family, seven to direct intervention by friends, eleven to financial problems created by drinking, seven to alcohol-related death or illness of another person, four to alcohol-related legal problems, and fifteen to extraordinary events, including personal humiliation, exposure to negative role models, events during pregnancy, attempted suicide, and personal identity crises. Many alcoholics listed more than one reason for the initiation of their recovery.

5. This conclusion would probably be contested by many strict behaviorists and sociologists whose primary emphases are more on environmental contingencies or events than ineffable psychological processes as determinants of behavior. Unfortunately, this is not a matter that can be easily resolved by logic or scientific "proof," since it relates to philosophical bias. My own bias is that an individual's thoughts and feelings play an important (but not always exclusive) role in how he or she subsequently will behave.

6. I have adopted this definition of hitting bottom from Greil and Rudy ("Conversion to the world view"). In an excellent commentary on the recovery process, Litman ("Personal meanings") describes these crisis points or "critical perceptual shifts" as times when "the individual is confronted with the choice of either a drastic change in drinking habits and lifestyle, or self-destruction."

7. In a recent study I have had an opportunity to document the reasons for 228 episodes of abstinence as reported by 150 alcoholics, 45 percent of whom were hospitalized. Details about the sample characteristics and interview procedure can be found elsewhere (Ludwig, "Pavlov's 'bells' "; and Ludwig, "Cognitive processes").

8. In my article, "Cognitive processes," I reported that three of twenty-nine respondents stopped drinking because they developed an allergy or physical aversion to alcohol. Here is an example.

Johnny had no real desire to quit drinking. He loved his scotch; he adored his beer. But then, years ago, he was treated by a psychiatrist with lithium for what was then diagnosed as a manic episode. While on this medication, he noticed that he would become "sicker than a dog" whenever he tried to drink alcohol. He remained on the lithium for some time, then decided to quit.

That Christmas, he experimented again with a highball. The next day he felt as though he had drunk two bottles of whiskey. His eye was swollen shut and he felt miserable. Months later, he tried to drink a beer and the same thing happened. It was almost with relief that he resolved never to drink again.

I have only been able to find an incidental reference to comparable allergies in the study by Gerard, Saenger, and Wile ("The abstinent alcoholic"). Despite the apparent low incidence of these allergies, their implications are intriguing, particularly if their authenticity can be confirmed, since they indicate that subtle biochemical changes, which are similar to those produced by a disulfiram-alcohol reaction, can protect individuals against the euphoric properties of alcohol. The clinical task is to identify these changes and then develop methods for inducing them in motivated individuals.

9. See the account by Shecky Greene in Wholey, *The Courage to Change*.

10. Note that in Lemere's retrospective study ("What happens to alcoholics") of 500 alcoholics, 22 percent of the patients quit drinking because of a potentially terminal illness, representing two-thirds of all those who became abstinent, a much higher percentage than in my study.

11. See *Alcoholics Anonymous*.

12. See *Alcoholics Anonymous* for accounts of individuals who managed to obtain sobriety without hitting bottom.

13. See Keller, "The oddities of alcoholics," for a delightful summary and interpretation of results of the numerous personality studies performed on alcoholics.

14. This is an extremely important issue. Unfortunately, many of the personality studies done on alcoholics have not adequately distinguished between "trait" and "state"—namely, the core, stable personality attributes preceding addiction or associated with full recovery and those superimposed attributes that are confounded by extreme reliance on alcohol or the physiological and psychological instability occurring during extended withdrawal from the drug.

15. See Tiebout ("The act of surrender"). Some of the terminology used by Dr. Tiebout seems dated and imprecise, but the general clinical observations are still remarkably perceptive. The term "defiance" should probably be supplanted by "denial," and "grandiosity" by "narcissism" to conform more to current designations for these ego mechanisms of defense.

16. See Bean ("Alcoholics Anonymous") for an excellent description of the psychodynamics of alcoholics prior to joining AA.

17. In his article, "On the loss-of-control phenomenon," Keller makes the point that the inability of alcoholics to refrain from starting drinking again is the hallmark of loss of control and a central feature of alcoholism.

18. This state of mind has been referred to as "deflation at depth." According to Bill Wilson ("Alcoholics Anonymous—beginnings and growth"), Dr.

Carl Jung was the first to employ this term when he pronounced Mr. R.'s condition utterly hopeless. This pronouncement "struck him at great depth, producing an immense deflation of his ego." Kurtz (*Not-God*) claims that deflation at depth represents a cornerstone and prerequisite for self-realization with the AA perspective.

19. See *Twelve Steps*.

20. See Bill Wilson ("Basic concepts"). This description is remarkable since it pertains as well to the states of mind associated with almost any radical transformation in attitudes and thought processes. Prior to almost any type of ideological or religious conversion, individuals also typically suffer from a sense of sin, a feeling of estrangement from God, guilt-laden depression, apprehension, sleeplessness, confusion, a sense of utter helplessness, and loss of personal pride. See James (*The Varieties of Religious Experience*), Lifton (*Thought Reform*), Starbuck (*The Psychology of Religion*), and Salzman ("The psychology of conversion").

Although the term "conversion" is not widely used in AA, it seems particularly apt for describing the major shift in personality manifestations and outlooks found in recovering alcoholics (see Tiebout, "Conversion as a phenomenon"). Salzman ("The psychology of conversion") adopts a broad definition of conversion, regarding it to be "any change in religion or of moral, political, ethical, or esthetic views which occurs in the life of a person either with or without a mystical experience, and which is motivated by strong pressures within the person."

21. Tiebout ("The act of surrender"), for whom "surrender" is an essential prerequisite for a successful recovery, makes an important distinction between this phenomenon and that of "submission," a far less desirable state of mind. "In submission, an individual accepts reality consciously but not unconsciously. He accepts as a practical fact that he cannot at that moment lick reality, but lurking in his unconscious is the feeling that 'there'll come a day' which implies no real acceptance and demonstrates conclusively that the struggle is still on. With submission, which is at best a superficial yielding, tension continues." On the other hand, with surrender, a process akin to conversion, the individual completely forfeits his narcissism and personal will and accepts the reality of his condition. As a result, a sense of relaxation, serenity, relatedness, freedom from conflict, and a newly found optimism ensue.

While the importance of surrender is generally acknowledged as a critical factor in recovery, like most generalizations it may not be appropriate in every single case. Jean Kirkpatrick (*Goodbye Hangovers, Hello Life*), for example, claims that many women neeed to develop more personal pride and self-respect as a basis for recovery rather than a greater sense of humility. Because of this, the AA emphasis on surrender of personal will to a Higher Power may not be suitable for them.

22. *Alcoholics Anonymous*, p. 27.

23. See Litman ("Personal meanings").

24. Litman ("Personal meanings") makes this very apt observation.

25. The observation that recovered alcoholics often switch from a value system centered around drinking to one centered around sobriety should not prove surprising, since it conforms well to Hoffer's thesis *(The True Believer)* that extremists of any kind find it far easier to switch from one extreme to another than to adopt positions of moderation. Militant atheists are more likely to become religious fanatics than to become agnostics. Except for the content of their beliefs, the radical right and the radical left tend to be just as militant and dogmatic in their views and tactics. Before World War II, dedicated communists in Germany were more likely to become ardent Nazis than to become moderates. While extremists of different persuasions tend to be implacable enemies, they also are potential soulmates; they accord each other a hatred or a grudging respect. Moderates appear on an entirely different wave length. Whereas moderates often adopt a balanced, middle-of-the-road approach to a particular issue, are amenable to compromise, and show tolerance toward a wide range of views, extremists are inflexible about their position, avoid compromise, and are intolerant toward all differing views. For extremists, particularly those with strong beliefs, there are only blacks and whites, no grays.

This observation, if valid, suggests one reason why most alcoholics are unsuccessful at controlled or social drinking. The alcoholic, by definition, is an extremist in his drinking and often centers his entire existence around alcohol. With the exception of the binge drinker, the thought of drink dominates his mind to different degrees morning, noon, and night, day after day. For an alcoholic to shift to a more moderate pattern of social drinking, where he has to weigh, decide, or choose how much and when to drink each day, whether an extra drink will really hurt or whether he has the ability to control his drinking in the future, represents a far more tenuous and shaky position than the extreme one of absolute abstinence with a lifestyle centered around staying sober. It is much simpler for the alcoholic not to have to deal with any ifs, ands, or buts, the qualifiers of a moderate. All he need do is avoid the first drink.

26. William Sargant, an English psychiatrist, in his book *Battle for the Mind*, proposes an intriguing neurophysiological explanation for mental changes of this sort. His speculations about what transpires during religious and ideological conversions, various healing ceremonies, thought reform, and the elicitation of confessions, which derive entirely from Pavlov's experiments with

animals, has direct relevance to the nadir or spiritual experiences also noted in alcoholics.

According to Sargant, Ivan P. Pavlov found that when the nervous system of certain types of dogs was strained beyond the limits of normal endurance, a rupture of nervous activity occurred, similar to what presumably happens in humans who break down under stress, along with a state of "transmarginal inhibition." Though the animal's behavior seems to deteriorate, this transmarginal inhibition serves a protective function for the brain, rendering it relatively immune or unresponsive to the originally disturbing stimuli and keeping the animal from acting in even more haphazard or bizarre ways.

After conditioning the dogs to a standard warning stimulus presented before the appearance of food, Pavlov employed four major types of stresses to produce mental breakdowns. The amount of saliva secreted by his dogs in response to the anticipation of food, compared to that amount secreted under normal conditions, represented a measure of how much transmarginal inhibition existed at any given time.

No doubt Pavlov exposed his animals to rather drastic and cruel procedures, but they are probably no worse than what many alcoholics inflict upon themselves over the years. The first type of stress he used was a steady increase in the intensity of a conditioned warning stimulus, such as an electric shock. Dogs continued to get their food but only after exposure to increasing pain. The second type of stress was the progressive prolongation of the intervals between the warning signal and the arrival of food. Signs of agitation and unrest usually began in the more unstable dogs as the waiting intervals got longer and longer. The third type of stress was the alternation of certain characteristics of the conditioned warning signals in order to confuse the dogs. Stimuli with positive, negative, or neutral valence would be presented unpredictably to heighten uncertainty and to keep the animals off balance. The fourth type of stress was the induction of physical exhaustion, glandular dysfunction, fevers, or gastrointestinal disturbances in order to weaken or dehydrate the dogs physically.

Pavlov described three progressive phases or types of protective transmarginal inhibition as the mental breakdowns began to take place. As with Pavlov's dogs, the ability of alcoholics to resist complete emotional collapse in response to persistent stress and, consequently, to display the comparable changes noted can be presumed to vary according to their basic temperaments, the state of their physical health, and the functioning of their nervous systems.

In the first or "equivalent" phase, the brain gives the same response to all stimuli, regardless of their strength. Weak and strong stimuli produce the same amount of saliva before the appearance of food. This phase has certain parallels to what transpires in alcoholics who, consumed with self-pity or indignation, display similar, blunted emotional responses to situations that once previously might have evoked joy or sorrow. The visit of a once-beloved daughter, for example, might arouse as much emotion as an interchange with a relative stranger.

With even greater stress, the equivalent phase of transmarginal inhibition gives way to a "paradoxical" phase in which the brain begins to respond more actively to weak than to strong stimuli. The dog, for example, will not accept food after a stronger stimulus but will after a weaker. This protective response against strong stimuli does not seem so different from the relative nonreactivity of certain alcoholics who are beset with one disaster after another—divorce, financial ruin, poor health—but who continue to overrespond to the petty irritations that plague their daily lives.

And then finally, with more persistent stress, the "ultraparadoxical" phase of transmarginal inhibition occurs, during which conditioned responses and behaviors suddenly switch from positive to negative or from negative to positive. A dog, for example, might act fondly toward a laboratory attendant it formerly disliked or attack the master it previously loved. Its behavior would become exactly opposite to all its past conditioning. It is of particular interest that this ultraparadoxical phase of transmarginal inhibition bears so many resemblances to the state of mind preceding therapeutic surrender, the "sweetly reasonable" and "softened up" state of mind which Bill Wilson regards as being most conductive to recovery, the same state of mind that predisposes individuals to a dramatic reversal in attitudes, radical transformations in beliefs, and a virtual 180 degree change in certain behaviors.

Should the bottom experience in alcoholics and the ultraparadoxical phase of transmarginal inhibition then be held to be equivalent? It is difficult to say. At best, this is an intriguing possibility, a theory in search of proof.

27. See Thomsen, *Bill W.*

28. See Bill W.'s detailed account of his spiritual experience in *Alcoholics Anonymous Comes of Age* and *Alcoholics Anonymous*, p. 63.

29. Bill Wilson makes this claim in *Alcoholics Anonymous Comes of Age*, p. 316. I am not certain how these figures were obtained. Hard scientific data from a large representative sample of alcoholics are lacking.

30. See Starbuck, *The Psychology of Religion.*

31. Musgraves, *One More Time*, p. 131.

32. See Jackson, *Afire with Serenity.*

33. See *Alcoholics Anonymous Comes of Age.*

34. See Hughes, *The Man From Ida Grove.*

35. See Tiebout, "Therapeutic mechanisms of Alcoholics Anonymous."

36. See Christensen, "Religious conversion"; Starbuck, *The Psychology of Religion*; and James, *The Varieties of Religious Experience.*

37. See Bryant, *The Magic Bottle*, for an account of the dramatic disappearance of craving following a religious experience. A number of alcoholics I personally interviewed also report similar experiences.

38. James, *The Varieties of Religious Experience*, p. 225.

39. What, then, are some alternative explanations? In a little-known essay, Gregory Bateson ("Cybernetics of 'self' ") examines the religious experience from an epistemological standpoint and offers an interesting analysis of why the notion of a personal God is so essential to the process of recovery, particularly within the framework of AA doctrine (see also Brundage, "Gregory Bateson").

"If the style of an alcoholic's ordinary sobriety drives him to drink," Bateson argues, with what appears to be irrefutable logic, "then the style must contain error or pathology; and intoxication must provide some—at least subjective—correction of this error." Most attempts at sustained sobriety fail mainly because of a fatal flaw in the alcoholic's ordinary conception of sobriety, which makes an artificial distinction between mind and matter, between conscious will or "self" and the remainder of the personality—a distinction reminiscent of Cartesian dualism, with its division between mind and matter. The first two steps in AA presumably correct this conceptual flaw *not* by the actual acts of "surrender" but by what it signifies—that is, a change in epistemology—a change in the habitual assumptions or premises by which the alcoholic knows about himself, others, and the world. This change in premises is required for sobriety. *"If a man achieves or suffers change in premises which are deeply embedded in his mind,"* Bateson claims, *"the results ramify throughout his whole universe."*

In his thesis, Bateson goes on to describe two basic relationships, each with its own set of epistemological assumptions. With the *symmetrical relationships*, every behavior or effort on the part of one party for dominance is met by a corresponding behavior or effort on the part of the other. Each party regards itself on equal status with the other. With the *complementary relationship*, every behavior or effort on the part of one party is met by a complementary behavior or effort on the part of the other, resulting, for example, in dominance and submission or nurturance and dependency. In the case of the alcoholic, the basic defect in his epistemology is presumed to be "pride," the symmetrical polarization between himself and his compulsion to drink as well as between himself and his fellow man and God. Because of this self-destructive pride, he tends to emphasize the "I can" and to repudiate the "I cannot." "The exercise of personal will," Starbuck wrote *(The Psychology of Religion)*, "is to live in a region where the imperfect self is most emphasized." Bateson echoes this sentiment by astutely pointing out that pride paradoxically narrows the concept of self, placing what happens outside its scope. Pride places the compulsion to drink outside the self, as something to be resisted, personal will pitted against relentless force. As long as the alcoholic clings to these assumptions, he is obligated constantly to take risks and to test his self-control. "I won't let alcohol get the best of me," he tells himself. "I can handle it." This time, unlike prior times, he convinces himself that he can control his intake of alcohol, but he inevitably fails.

When the alcoholic eventually sobers up, the struggle of wills begins anew until another episode of drunkenness, and then the cycle repeats itself again and again until the entire system gets out of control and he "hits bottom," the point at which his epistemology of self-control becomes completely bankrupt. Since he *cannot not* drink, he is forced into a logic trap offering no apparent escape. An obsession of mind compels him to drink, even though he must not, but if he does, as Dr. W. D. Silkworth, one of the founders of Alcoholics Anonymous, reputedly said, he will experience an "allergy of body that condemns him to go mad or die" (see A.A. *Grapevine*, p. 7). In other words, his impossible choice is between going mad and dying or going mad and dying. The only possible way out of this trap, other than unsuccessfully trying to avoid it by intoxication or suicide, is for the alcoholic to experience a spontaneous or involuntary change in his basic epistemology, a shift from a symmetrical to a complementary relationship with God, such as happens during a sudden awakening or the gradual development of strong religious convictions over time.

Once the alcoholic entrusts himself to a Higher Power or personal God, a power which, unlike him, is not helpless in dealing with alcoholism (for many, AA itself may come to symbolize a godlike organization superordinate to the individual), he no longer has to engage alone in his struggle. He is now part of a greater whole, a hierarchy of mental systems which culminate in an ultimate Mind or God. As such, he no longer has a need to test his self-control periodically. And as long as he has no need to test himself, he should have no motivation to drink. So goes the epistemological analysis of sobriety.

But demonstrating the epistemological necessity of a personal God for the process of recovery offers no explanation of how individuals come to identify the source of their extraordinary experiences and the changes that follow as divine in nature. Why not attribute the religious experience to a wish-fulfilling hallucination, as in a dream or hypnogogic state, or to delirium tremens, alcohol intoxication, or altered brain chemistry, particularly when the experience presumably only arises in individuals recently inebriated, malnourished, dehydrated, sleep-deprived, or distraught? How do individuals make the distinction between a sacred and a profane cause of their experience?

Attribution theory, which deals with how people make causal inferences, offers potential answers to these questions. It represents a kind of "psychological epistemology," since it deals with the processes by which people know their world and know that they know (see classic articles by Kelly on "Causal attribution" and "Attribution theory"). The theory maintains that individuals, for the most part, tend to make inferences about subjective causality and reality mainly on the basis of the covariation principle, whereby an effect is attributed to the one of its possible causes with which it varies over time. There are three main sets of causes: persons (i.e., the individual or others), entities (i.e., outside forces or objects), and times (i.e., situations or circumstances).

Attribution of the cause of an effect to the person himself (or oneself) presupposes that the person and no other acts in a particular way toward a given entity and similar entities at all times and circumstances. In the case of a spiritual experience, the cause would be attributed to the individual himself (or his peculiar state of mind) rather than a divine being only if that particular individual had such an experience under all kinds of conditions—that is, in situations both conducive and nonconducive to divine intervention and in different states of mind. Since other people report comparable spiritual experiences usually under a rather restricted set of conditions, such as delirium tremens or intoxication, this type of attribution seems clearly untenable.

Attribution of the cause of an effect to the particular time or circumstance presupposes that a given individual, and likely all individuals, respond in a similar way only during those times or circumstances, whether or not they have been exposed to different kinds of outside forces. In the case of a spiritual experience, the cause would be attributed to the special circumstance (e.g., low point, delirium tremens, intoxication) rather than to the alcoholic himself or some outside entity if that particular individual or all others like him experience a divine presence only during these particular circumstances. This type of causal inference can be invoked to account for most reported spiritual experiences, which tend to occur only under these very special circumstances.

Attribution of the cause of an effect to an outside entity (or God) presupposes that a given individual, and likely all individuals, respond only to that particular entity in a certain manner at all times and under all circumstances. Otherwise, it would not be possible to determine whether the response represents certain properties of the particular individual or the particular circumstances. In the case of a spiritual experience, the cause could be attributed to a personal God or outside force if that individual and all similar kinds of individuals reacted in a certain manner at all times and circumstances whenever that entity makes itself known. The fact that many alcoholics have comparable religious experiences primarily when they are at a low point in their lives strengthens this kind of attribution; the fact that countless other alcoholics do not and that these experiences tend to happen only under very restricted conditions tends to weaken this kind of causal inference. Therefore, from the standpoint of strict attribution theory, it is difficult to make a logically airtight case for the intervention of a personal God.

While attribution theory instructs us on how inferences of subjective causality tend to be made, it offers only limited aid about how individuals should interpret the nature of their spiritual experiences. That is because of all of the confounding factors associated with the individual's spiritual experience or religious conviction in divine intervention. The observation that these spiritual experiences are not available to all alcoholics when they are in extremis, that the nature of the experience differs greatly among individuals, varying from transcendental and mystical to discretely hallucinatory in character, and that they apparently occur under highly restricted conditions, leaves the conceptual door open for individuals to make almost any inference they choose.

In the case of individual alcoholics, who likely have little interest in these theoretical matters, causal inferences and subsequent convictions must always be leaps of faith. Analytical logic and scientific information-gathering are completely foreign and inappropriate methods for assessing the authenticity and personal significance of their spiritual experiences. The issue for them is relatively simple. Who or what else, they ask themselves, but a supreme being could induce so profound an experience, rescue them from the abyss of despair, the horrible trap of their situation, and account for such truly miraculous changes? There are, however, other possible explanations that bear on attribution theory. But before discussing this, here are some background considerations.

The classic study by Stanley Schachter and James E. Singer ("Determinants of emotional state") on the nature of emotions was devised to resolve much of the controversy generated by the James-Lange theory of emotions, which held that the experience of a particular emotion depended upon the specific nature of the underlying physiological changes, and Walter Cannon's later criticisms that the same visceral changes occur in very different emotional states as well as in nonemotional states. Schachter and others later attempted to break this conceptual stalemate by proposing that cognitive factors were probably the major determinants of emotional states and that individuals labeled, interpreted, and identified their aroused visceral states on the basis of their cognitive expectations and evaluative needs. The identification of a specific emotion, as fear, euphoria, or awe, should depend not only on the stimulation of a general and diffuse pattern of physiological excitement, such as might be produced by an injection of adrenaline, but also on the attribution or cognitive label or explanation provided by the individual to account for this change.

For the study in question, which has important implications for the understanding of spiritual experiences, the experimenters gave subjects either disguised injections of adrenalin or a placebo while they supposedly awaited a "vision test." Those receiving adrenalin were either informed, misinformed, or kept ignorant about what physiological effects to expect from the injections. Subsequently, they were exposed to one of two conditions. In the euphoria condition, an accomplice, who also was brought into the room, began acting silly and playful; in the anger condition, he began acting querulous, nasty, and sullen.

The results were revealing. Given a state of physiological arousal for which individuals had a completely satisfactory explanation, they did not label this state in terms of the alternative cognitions available. When subjects were told exactly what they would feel and why after an injection of adrenalin, they neither reported the anger nor euphoria other subjects reported who had no knowledge of why they felt the way they did. When individuals did not have immediate, satisfactory explanations for how they felt, then the same state of physiological arousal could give rise to a diversity of emotional labels, including euphoria or anger, depending upon the cognitive demands of the situation. In the absence of any physiological arousal, individuals were not likely to identify their feelings as specific emotions, regardless of the cognitive circumstances.

The Good Friday study of Walter Pahnke ("Contributions of religion to the therapeutic use of psychedelic substances" and "Implications of LSD") sheds further light on these results. Pahnke administered either psilocybin, an hallucinogenic substance extracted from "magic mushrooms," or nicotinic acid, a vitamin that produces warmth and tingling of the skin but has no mental effect, to twenty divinity students on Good Friday. Then, inside a private chapel, they listened to a 2½-hour religious service consisting of organ music, four solos, readings, prayers, and personal meditation. Compared to those who received the placebo, the divinity students who received psilocybin were much more apt to report mystical experiences characterized by a sense of undifferentiated unity, the blurring of boundaries between their selves and the outside world, the sense of illumination and certainty about one's insights and perceptions, the transcendence of space and time, a sense of sacredness, awe, or reverence, a feeling of joy or peace, the suspension of their logical faculties, a sense of ineffability and transiency about the experience, and, most important, a subsequent positive change in attitude toward themselves, others, life, and even God.

While Pahnke appropriately emphasized that the *mental set* of the students (i.e., their anticipation of a religious experience on Good Friday) and the actual physical *setting* (i.e., the chapel) represented crucial factors in transforming the undifferentiated, hallucinogenic state into a mystical experience, we should not overlook the inherent potential of the hallucinogenic drug itself for stimulating the religious sentiment or expanding consciousness, regardless of the expectations of individuals.

Aldous Huxley *(The Doors of Perception* and *Heaven and Hell),* after his personal experience with mescaline, claimed that this agent, as well as alcohol, had the capacity to heighten religious sentiment by enabling individuals to shake off their egos and experience the glorious feeling of release, as once happened with those who worshipped Dionysus, the god of wine, or the Celtic god of beer. The North American Indians have employed peyote ritually during ceremonies in the Native American Church. (La Barre, *The Peyote Cult,* states that there is no scientific evidence to support the claim that the ritualized use of peyote protects Indians against the ravages of alcoholism.) Taking issue with Huxley's position, Zaehner *(Mysticism)* argues that these drugs induce an artificial or false sense of the religious sentiment, and desecrate the true mystical experience.

From numerous studies, it is apparent that these drugs can and do stimulate unusual mental experiences. Mystical, religious, or spiritual experiences have also been reported after prolonged sensory deprivation, in dreams, trance states, meditation, severe emotional turmoil, auras preceding seizures, fever, and from the administration of many psychotropic drugs, such as nitrous oxide, opiates, cocaine, amphetamines, volatile hydrocarbons, barbiturates, and even alcohol. They also have been noted during alcohol intoxication and alcohol withdrawal. What seems to be common to all these conditions is an altered state of consciousness, a nonrational state of mind in which a heightened sense of significance and other-worldly and unusual experiences abound (see Ludwig, "Altered states of consciousness"). Usually, in these unsettled states, when ordinary reality props appear to be dissolving away, individuals are much more susceptible to outside suggestion or their own personal needs. This is demonstrated most clearly in our own studies years ago (Ludwig and Lyle, "Narcotic drug effects through hypnosis"). Hypnotized narcotic addicts, for example, when falsely told that they were getting intravenous salt water, showed no subjective or behavioral effects of the actual morphine injections, while those who were told that they received morphine injections demonstrated most of the subjective, behavioral, and even pharmacological effects associated with "mainlining" this drug, even though they actually received saline. In later treatment studies, first with narcotic addicts (Ludwig and Levine, "LSD, hypnosis, and psychotherapy") and then with alcoholics (Ludwig, Levine and Stark, *LSD and Alcoholism),* we found that their experience with lysergic acid diethylamide (LSD), a potent hallucinogen, was very malleable and could be substantially, but not entirely, modified either by hypnotic suggestion or the constraints of the therapeutic setting, results which were comparable to those obtained in the Good Friday experiment.

How do all these observations pertain to the spiritual experiences of alcoholics? Possibly not at all, possibly very much. The extent to which they do would suggest that a state of nonspecific, physiological arousal, such as can be experienced during relative intoxication, mild delirium, sleep deprivation, excessive fatigue, or extreme emotional distress, is required initially as a means of stimulating the evaluative needs of individuals so that they can better explain what is happening to them. The transitory auditory and visual illusions, the minor hallucinatory phenomena, the mild seizure discharges, a vivid waking dream or some unknown brain state, so common with these conditions, can provide this arousal. In the absence of ready, plausible explanations for all the confusing and ambiguous changes within, individuals are apt to see, hear, or experience whatever they must—whatever conforms to powerful inner needs and strong, outer expectations—to escape from their impossible, epistemological traps. For those who know through personal experience or have heard from testimonials at AA meetings or from other sources that salvation can come from a spiritual experience, what better way to interpret their unusual experiences than as the result of divine intervention? What a relief it is to escape the boundaries of self, to achieve the joy of at-oneness, the sense of communion so common to altered states of consciousness. And along with that comes a heightened sense of truth, meaning, and significance in what they now perceive, think, and feel (see Bowker, *The Sense of God*; and Proudfoot and Shaver, "Attribution theory," for a discussion of the importance of cognitive labeling or attributions for the interpretation of religious experiences. The latter authors regard attribution theory as "a natural path into the realm of religion," p. 325).

Because it contributes so directly to the unshakable conviction in the reality of the experience, this heightened sense of truth, meaning, and significance, which is inherent to all revelatory states, requires special comment. The point that needs to be made is that the *sense* of meaning or truth cannot be regarded as equivalent to *real* meaning or truth—if that can ever be established. This distinction was first brought home to me during a personal experience long ago when I took LSD for experimental purposes. Sometime during the height of the reaction, I had an intense desire to urinate and made my way on rubbery legs to the bathroom, escorted by my colleague. As I stood by the urinal, marveling at my elongated neck and the exquisite sense of relief, I suddenly became aware of a sign whose words filled me with rapture. The more I weighed those words, the more I realized their profound meaning. No wonder philosophers throughout the ages had been unable to discover the meaning of life. The answer was obvious. What they failed to comprehend was that the truth was right before their eyes. Thrilled by this revelation, I rushed back to my colleague to share this incredible insight with him. With my fly undone and a maniacal gleam in my eye, he must have taken me for a madman since he resisted my efforts to pull him into the bathroom. Finally, when I managed to do so, he burst out laughing. Being a mere mortal, he could not appreciate the world-shaking import of the revelatory sign, which read, "Please Flush After Using."

Sometime later, I was invited to a college to talk about my studies with drugs and mentioned this experience to a group of students to illustrate that the subjective truth that people experience under drugs need not have any bearing on objective truth, when one of the students asked me how I could be certain that the words of the sign, in fact, did not convey the true meaning of life. The audience laughed and I did too, but I was thankful that I was not pressed to answer the question then, since at that time I could not have given a worthy, philosophical response. Since then, I am better prepared (see Ludwig, "Characteristics of therapeutic insight").

What I had experienced personally, William James had noted much earlier for alcohol and other chemicals. "One of the charms of drunkenness," he writes in his *Outline of Psychology*, "unquestionably lies in the deepening sense of reality and truth which is gained therein." In *The Varieties of Religious Experience* (p. 378), he adds:

"Nitrous oxide and ether, especially nitrous oxide, when sufficiently diluted with air, stimulate the mystical consciousness in an extraordinary degree. Depth upon depth of truth seems revealed to the inhaler. This truth fades out, however, or escapes, at a moment of coming to; and if the words remain over in which it seemed to clothe itself they prove to be the veriest nonsense. Nevertheless, the sense of a profound meaning having been there persists; and I know more than one person who is persuaded that in the nitrous oxide trance we have a genuine metaphysical revelation."

Unfortunately, what transpires during an alteration in consciousness or unstable state of mind, whether or not of a revelatory nature, cannot adequately explain many of the remarkable mental changes that may take place subsequently, the transformation from an alcoholic to a nonalcoholic world view. Attribution theory and epistemology may shed some light on how these experiences are interpreted, but they leave many questions unanswered. What possible mechanisms can account for these dramatic mental rearrangements, this recrystallization of values, these shifts in a person's psychological "center of gravity" that bring, in William James' phraseology, dead feelings, ideas, and beliefs to life and lay to rest the old ones (see *The Varieties of Religious Experience*). Fortunately, there are ways to conceptualize what happens. Later, we shall consider how such transformations may come about.

40. Christensen ("Religious conversion") lists some possible outcomes for a conversion experience: the reintegration may resolve the conflict and permit the ego to reintegrate at a more optimum level of functioning than before; it may take place at the previous level; it may take place at a lower level; or

rarely, it may not take place at all, especially if it is part of a psychotic or psychologically unstable process.

That an abrupt religious conversion experience need not lead to a stable and lasting adjustment afterward is illustrated by Starbuck's study (see *The Psychology of Religion*) of 100 evangelical church members, which revealed that there had been backsliding of some nature within two weeks in 93 percent of the women and 75 percent of the men, mostly with respect to their enthusiasm and faith.

In a more anecdotal manner, Bryant (*The Magic Bottle*, p. 149), for example, describes the situation of a friend who recently claimed to be saved. "The life Isabel had in Christ seemed as transparent, as insubstantial, as that of a jelly fish. She oscillated hourly between anguish and exhilaration, multiplying the grandeur of her victories and exaggerating psychologically the seriousness of her defeats." It is difficult to know how many other alcoholics find themselves in a comparable situation.

Chapter 7

1. I am indebted to M. C. Maultsby, Jr., for first drawing my attention to differences in cognitive style between individuals who think predominantly in images and those who think predominantly in concepts.

2. G. A. Marlatt (see Marlatt and Gordon, *Relapse Prevention*) attempts to train his alcoholic patients to recognize the "apparently irrelevant decisions" as early warning signs of an impending "slip" and to take appropriate preventive steps.

3. G. A. Marlatt (see Marlatt and Gordon, *Relapse Prevention*) appropriately emphasizes the importance of "detachment" as an effective means of coping with cravings.

4. I am indebted to M. C. Maultsby, Jr., (*Rational Behavior Therapy*) for his perceptive analysis of this matter.

5. Litman ("Relapse in alcoholism") lists distraction as one of the four main types of coping behaviors in alcoholic "survivors." The other behaviors include thinking positively about sobriety, negative thoughts about drinking, and avoidance of drinking situations.

6. In psychoanalytic theory, the belief in the thought as a reality, as an integral part of what an individual truly feels and wants to do, is commonly found in children, primitive societies, and psychotic individuals. It is also found widely in society in general, particularly in individuals exposed to more fundamentalist, religious beliefs in which the thought may be regarded as

equivalent with the deed. With "sins of attrition," an individual is just as responsible for a bad thought as a bad behavior.

7. G. A. Marlatt (see Marlatt and Gordon, *Relapse Prevention*) has hit upon a graphic variant of this technique which he routinely employs in his relapse prevention program. Alcoholics are told to picture the urge to drink, when it strikes them, as a wave which surges and then abates as they ride its crest like surfers into shore. Another image he employs is to have the alcoholic imagine himself as a samurai warrior beheading an enemy representing a symbol of the desire to drink. As clever as these metaphorical images are, almost any ones will do, provided that the individual gradually dissipates the power of the impulse and symbolically renders it impotent.

8. In my study (see Ludwig, "Cognitive processes"), roughly 48 percent of recovered alcoholics developed automatic negative images to the thought of a drink. A substantial number of sober alcoholics also reported using this technique in the study by Litman et al. ("Differences in relapse precipitants").

9. See Cautela, "Covert sensitization."

10. In a most interesting study, Elkins ("Covert sensitization") and Elkins and Murdoch ("The contribution of successful conditioning to abstinence maintenance") demonstrated that those alcoholics who achieved automatic conditioned nausea to the thought of drinking displayed a higher abstinence rate on follow-up than those who conjured up the experience of nausea only when asked to and a control group in which this procedure was not applied.

11. These techniques are comparable to those effectively used by Recovery, Inc., a self-help organization for mental patients (see Low, *Mental Health Through Will Training*; Wechsler, "The self-help organization").

12. Many of these same approaches have been used with some degree of success as behavior modification treatments for obsessions, compulsions, and phobias and are largely derived from learning theory principles pertaining to reinforcement and extinction (see Emmelkamp, *Phobic and Obsessive-Compulsive Disorders*; Rachman and Hodgson, *Obsessions and Compulsions*.)

13. I am indebted to De Silva ("Early Buddhist") as the source for my description below of the techniques employed by Buddhist monks to deal with unwanted thoughts.

14. Litman ("Relapse in alcoholism") reports that decreased "cognitive vigilance" represented one of the main precipitants to relapse in her sample of alcoholics.

Chapter 8

1. See Goodwin, "The alcoholism of Eugene O'Neill."

2. For an excellent description of what is meant by a "state of mind," see

Horowitz *(States of Mind)*. His model, if I understand it correctly, seems ideally suited to explain various aspects of the transition from a drinking-oriented frame of mind to one of sobriety. I recommend this book highly for the reader interested in learning more about the factors that contribute to psychological stability and change.

3. Various cognitive-behavioral theorists have provided far more sophisticated analyses of how certain types of thought processes affect human behavior (see Note 2 in Chapter 4 for a selected listing of authors).

4. This notion of mental scripts is not original. In his book, *A Million Dollars for Your Hangover*, Maultsby gives examples of certain attitudes conducive to drinking. He also anecdotally describes having an alcoholic patient listen to a tape recorded by the patient himself of a nondrinking script that offers rational reasons not to drink while picking up and smelling his favorite drink, but then sipping a glass of juice instead.

5. See Ludwig, "Cognitive processes"; and Ludwig, "Pavlov's 'bells' " for further details of this study.

6. Not all alcoholics, unfortunately, are able to program their own minds with appropriate sobriety scripts. For many of these individuals, frequent attendance at Alcoholics Anonymous meetings and reading appropriate material may provide sufficient, credible, and convincing scripts to sustain their abstinence.

7. In his book, *The American Alcoholic*, William Madsen offers a Jamesian conceptualization on how and why this switch from one set of values to another takes place. His thesis is that all people yearn for simplicity. As societies become more civilized, individual values tend to multiply and drift from a core of universals into the sphere of alternatives, where values and propositions often conflict and collide. In more primitive communities, where life is uncomplicated and values are stable, people are apt to know exactly what they believe and exactly where they stand, not having been exposed to as much diversity of opinions and beliefs as those in modern societies. Because the existence of alternative values raises anxiety, mainly because they challenge accepted beliefs, individuals prefer simplification, seeking a new system that transforms these conflicting values to integrated ones. Drunkenness and sobriety are a case in point. An individual cannot comfortably believe in and express both simultaneously. Though he may not be able to accomplish it, he has a need to commit himself to one or the other mainly because if he can't, he will never know inner peace. He will become guilty when he drinks because he knows he shouldn't, and become anxious when he abstains because he knows he doesn't want to. It is this desire for simplification that eventually induces him to adopt one integrated value system or another.

Chapter 9

1. Redgrove, *The Beekeepers*.

2. Most treatment studies reveal a highly significant relationship between drinking behavior and other parameters of general functioning. In our follow-up study (Ludwig, Levine, and Stark, *LSD and Alcoholism*), the results indicated that the less alcoholics drank, the better their health, employment situation, legal status, and social adjustment.

3. See also *Alcoholics Anonymous* for detailed accounts of individuals who have attained sobriety.

4. My own formula for sobriety, as determined from my studies of alcoholics, shares many elements with those of other investigators in the field. Given the importance of this topic, what is particularly astounding is that more scientific research and thought are not available for consideration.

From her interviews with alcoholic "survivors," that is, those who managed to remain sober for variable lengths of time after participation in a treatment program, Gloria K. Litman, an English psychologist ("Personal meanings"), describes a sequence of steps *en route* to sobriety. The initial step is a "critical perceptual shift," a dramatic change of self-concept and personal expectations that happens when individuals, as a consequence of hitting bottom, feeling helpless, or being profoundly distressed, have their habitual defences of denial, distortion, and projection shattered. What typically occurs is that individuals who previously sought out reasons to drink, failed to utilize feedback from family members and friends about the destructive aspects of their drinking, and were relatively immune to therapeutic interventions, now adopt entirely different orientations and values. Because of their feelings of desperation and helplessness during this critical time, individuals not only become more open to help but also tend to attribute magical or extraordinary qualities to the helper, institution, or treatment program—their only lifeline to sobriety or salvation—which serves *in loco parentis* or *in loco dei* for them. Within the context of this relationship, recovering alcoholics begin to make a commitment about restructuring their lives. Old values are abandoned and new ones adopted. Individuals become aware of the destructive, self-defeating nature of their attitudes and behaviors, and try to change them accordingly. Sober-mindedness rather than drunkenness or intoxication becomes their overriding goal. In order to accomplish and sustain these remarkable changes, individuals adopt a number of short-term strategies to protect themselves against relapse, particularly during their most vulnerable time, the first few months after they have stopped drinking. One common tactic is avoidance of places and times in which alcohol is available. Another is learning how to refuse a drink. Another is the practice of negative thinking, whereby the images or thoughts of alcohol come to be associated with undesirable consequences. Another involves the restructuring of free time, filling the void of what formerly had been occasions to drink with opportunities for constructive or enjoyable activities. Changing the social environment from one conducive to drinking to one promoting the pleasures of clearheaded thought and conversation is yet another tactic, as is short-term problem solving, dealing with the unresolved aftermath of past transgressions and the nagging problems surrounding a new start. And another involves engaging in therapy or affiliation with a helping agency in order to derive the benefits of group support and constructive feedback from others. As recovering alcoholics begin to gain confidence in their newfound abstinence, their goals now begin to become more ambitious and far-reaching. Their time perspective increases. They can start to think beyond the immediate survival strategy of getting by each hour or day and can contemplate more long-range goals. They are less exploitative of others. Their problem-solving skills expand. Their self-esteem continues to grow as positive feedback increases. And with this heightened self-esteem, they engage in much more positive thinking, whereby sobriety and the lifestyle associated with it become much more highly valued than ever before. Now that sobriety represents a more realizable goal, individuals can start making reparations to those who have suffered as a result of their past excesses and begin rebuilding their lives on a more solid footing. They are off "cloud nine" now, so their aspirations are more feasible and realistic. And with an increase in their abilities to tolerate frustration, they become far better able to engage in long-term problem solving and to lead more independent lives without relying as much on outside helping agencies. They have achieved sobriety.

In distinction, the model developed by B. S. Tuchfeld ("Spontaneous remission"; and, with S. H. Marcus, "The resolution of alcohol-related problems") has a somewhat different emphasis. After recognition of his or her problem, which may be triggered by deterioration in health, occupational problems, legal encounters, a religious experience, or identity crises, the individual presumably makes a commitment to a particular course of action. The strength of this commitment is dependent on a number of *moderator* variables, such as willingness to identify him- or herself as an alcoholic, receptivity to help, and motivation for change. Once this commitment is made, its continuation is dependent upon a number of *maintenance* factors, which are both social and psychological in nature. Social factors pertain to the availability of nonalcohol-related leisure and social activities, positive reinforcement from family and friends, a stable economic situation, and marital and social support. In many instances, a change in significant others, such as falling in love, getting married, or joining a group with strong antidrinking attitudes, may facilitate the quest for abstinence.

From a psychological standpoint, the individual is helped by "justifying rhetorics" adopted to explain past behavior. Individuals have a great intolerance for the unknown and seek to make sense out of what happens to them, particularly when what happens is discordant with their image of themselves as responsible, decent human beings. These explanations, which may come from religion, AA, psychotherapy, or other explanatory systems, serve to provide a more stable conceptual foundation upon which individuals may build a rationale for their newfound sobriety.

As health, economic, and other benefits accrue from these changes in attitudes and behavior, individuals are encouraged and stimulated to continue in this constructive direction. Sobriety, in a sense, becomes its own reward and reinforces the commitment for this lifestyle.

Though not explicitly stated, certain features of G. A. Marlatt's relapse prevention program ("Cognitive assessment and intervention") appear to parallel the process of recovery. Employing the metaphor of a journey, Marlatt indicates that alcoholics must pass through a number of stages in order to accomplish habit change. Preparation for departure requires a strong sense of motivation and commitment for abstinence. During this time, individuals must become aware of the high-risk situations and learn how to recognize trouble spots that predispose to relapse. They try to cope with their urges and cravings "one day at a time." An important aspect of coping involves learning to identify the early warning signals for relapse and the "apparently irrelevant decisions" secretly designed to encourage "slips." The journey itself, once undertaken, is facilitated by the enlistment of social support from family, significant others, or possibly AA. In addition, certain behavioral techniques may aid the recovery process. These techniques include the avoidance of situations associated with prior heavy drinking and the employment of substitute activities, such as sipping on soft drinks or carrying a rubbing stone in lieu of consuming or handling liquor. So that sobriety be likewise associated with pleasure, individuals must make an effort to engage in a variety of pleasurable activities, such as going out for supper and the movies, taking trips, or attending sporting events.

While on this journey, individuals can periodically expect to experience strong temptations, but they must develop a sense of detachment about their urges, relieving themselves of the necessity of acting upon them. When avoidance, distraction, and other coping techniques fail to work, individuals must be prepared to deal with the potential of a slip. Forewarned is forearmed. An initial lapse need not inevitably lead to a full-blown relapse. Once back on the road to recovery, individuals need to find some way to bolster their motivation and recommit themselves to their difficult goal. Remaining hypervigilant about their drinking problem and exposing themselves to constructive confrontation by others, such as may happen in AA or therapeutic groups, represent ways in which this can be done.

5. See also *Living Sober*.

6. This conceptualization has been advanced by Glasser in his book, *Positive Addiction*.

7. The recommendation for "substitute indulgences" represents a formal part of the relapse prevention program developed by Marlatt (Marlatt and Gordon, *Relapse Prevention*).

8. A comparable list of activities can be found in *Living Sober*.

Bibliography

A.A. Grapevine, 7:12 (May), 1951.

Adams, R. L., and Fox, R. J., Mainlining Jesus: the new trip, Society, 9: 50–56, 1972.

Alcoholics Anonymous, New and Revised, 2nd Edition, Alcoholics Anonymous World Services Inc., New York, N.Y., 1955.

Alcoholics Anonymous Comes of Age: A Brief History of A.A., Alcoholics Anonymous World Services Inc., New York, N.Y., 1957.

Anonym, K., *Understanding the Recovering Alcoholic*, 2nd Edition, Hazelden Foundation, Center City, Minn., 1980.

Antze, P., The role of ideologies in peer psychotherapy organizations: some theoretical considerations and three case studies, J. Applied Behavioral Science, 12: 323–346, 1976.

Armor, D. J., Polich, J. M., and Stambul, H. B., *Alcoholism and Treatment*, The RAND Corporation, Santa Monica, Calif., 1976.

Babayan, E. A., and Gonopolsky, M. H., *Textbook on Alcoholism and Drug Abuse in the Soviet Union*, International Universities Press, New York, N.Y., 1985.

Baekeland, F., Evaluation of treatment methods in chronic alcoholism, in B. Kissin and H. Begleiter (eds.), *The Biology of Alcoholism, Vol. 5: Treatment and Rehabilitation of the Chronic Alcoholic*, Plenum Press, New York, N.Y., 1977.

Bailey, M. B., and Stewart, J., Normal drinking by persons reporting previous problem drinking, Quart. J. Stud. Alcohol, 28: 305–315, 1967.

Baker, T. B., and Cannon, D. S., Taste aversion therapy with alcoholics: techniques and evidence of a conditioned response, Behav. Res. & Therap., 17: 229–242, 1979.

Bandura, A., Self-efficacy: toward a unifying theory of behavior change, Psychological Review, 84, 191–215, 1977.

Bateson, G., The cybernetics of "self": a theory of alcoholism, in *Steps to an Ecology of Mind*, Ballantine Books, New York, N.Y., 1972.

Bateson, G., *Mind and Nature*, Dutton, New York, N.Y., 1979.

Bean, M., Alcoholics Anonymous, Psychiatric Annals reprint, No. 2 & 3, 1–64, 1975.

Beck, A. T., Rush, A. J., Shaw, B. F., and Emery, G., *Cognitive Therapy of Depression*, Guilford, New York, N.Y., 1979.

Beckman, L. J., An attributional analysis of Alcoholics Anonymous, J. Stud. Alcohol, 41:714–726, 1980.

Blane, H. T., Issues in the evaluation of alcoholism treatment, Professional Psychology, 8: 593–608, 1977.

Bowker, J., *The Sense of God*, Clarendon Press, Oxford, U.K., 1973.

Brickman, P., Rabinowitz, V. C., Karuza, J., Coates, D., Cohn, E., and Kidder, L., Models of helping and coping, American Psychologist, 37: 368–384, 1982.

Bruccoli, M. J., *Selected Letters of John O'Hara*, Random House, New York, N.Y., 1978.

Brundage, V., Gregory Bateson, Alcoholics Anonymous and Stoicism, Psychiatry, 48: 40–51, 1985.

Bryant, L., *The Magic Bottle*, A. J. Holman, Philadelphia, Pa., 1978.

Caddy, G. R., and Block, T., Behavioral treatment methods for alcoholism, in M. Galanter (ed.), *Recent Developments in Alcoholism*, Vol. 1, Plenum, New York, N.Y., 1983.

Cahalan, D., Cisin, I. H., and Crossley, H. M., *American Drinking Practices: A National Study of Drinking Behavior and Attitudes*, Rutgers Center for Alcohol Studies Publication Division, New Brunswick, N.J., 1969.

Cautela, J. R., The treatment of alcoholism by covert sensitization, Psychotherapy, 7: 86–90, 1970.

Christensen, C. W., Religious conversion, Arch. Gen. Psychiat., 9: 207–216, 1963.

Clancy, J., Vanderhoff, E., and Campbell, P., Evaluation of an aversive technique as a treatment of alcoholism: controlled trial with succinylcholine-induced apnea, Quart. J. Stud. Alcohol, 28: 476–485, 1967.

Cohen, M., Liebson, I. A., and Faillace, L. A., The modification of drinking of chronic alcoholics, in N. K. Mello and J. H. Mendelson (eds.), *Recent Advances in Studies of Alcoholism: an Interdisciplinary Symposium*, National Institute of Alcohol Abuse and Alcoholism, Rockville, Md., 1971.

Cooney, N. L., Baker, L. H., and Pomerleau, O. F., Cue exposure for relapse prevention in alcohol treatment, in K. D. Craig and P. J. McMahon (eds.), *Advances in Clinical Behavior Therapy*, Brunner/Mazel, New York, N.Y., 1983.

Cummings, C., Gordon, J., and Marlatt, G. A., Relapse: prevention and prediction, in W. R. Miller (ed.), *The Addictive Behaviors*, Pergamon Press, Oxford, U.K., 1980.

Curry, S. G., and Marlatt, G. A., Unaided quitter's strategies for coping with temptations to smoke, in S. Shiffman and T. A. Wills (eds.), *Coping and Substance Use*, Academic Press, New York, N.Y., 1985.

Davies, D. L., Normal drinking in recovered alcohol addicts, Quart. J. Stud. Alcohol, 23: 94–104, 1962.

De Silva, P., Early Buddhist and modern behavioral strategies for the control of unwanted intrusive cognitions, Psychological Record, 35: 437–443, 1985.

Donovan, D. M., and Chaney, E. F., Alcohol relapse prevention and intervention: models and methods, in G. A. Marlatt and J. R. Gordon (eds.), *Relapse Prevention*, Guilford Press, New York, N.Y., 1985.

Drew, L. R. H., Alcoholism as a self-limiting disease, Quart. J. Stud. Alcohol, 29: 956–965, 1968.

Edwards, G., Arif, A., and Hodgson, R. J., Nomenclature and classification of drug- and alcohol-related problems: a WHO memorandum, Bulletin World Health Organization, 59: 225–242, 1981.

Edwards, G., and Gross, M. M., Alcohol dependence: provisional description of a clinical syndrome, Brit. Med. J., 1: 1058–1061, 1976.

Edwards, G., Gross, M. M., Keller, M., and Moser, J., Alcohol-related problems in the disability perspective: a summary of the WHO group of investigators on criteria for identifying and classifying disabilities related to alcohol consumption, J. Stud. Alcohol, 37: 1360–1382, 1976.

Edwards, G., Orford, J., Egert, S., Guthrie, S., Hawker, A., Hensman, C., Mitcheson, M., Oppenheimer, E., and Taylor, C., Alcoholism: a controlled trial of 'treatment' and 'advice', J. Stud. Alcohol, 38: 1004–1031, 1977.

Elkins, R. L., Covert sensitization treatment of alcoholism: combinations of successful conditioning to subsequent abstinence maintenance, Addictive Behav., 5: 67–89, 1980.

Elkins, R. L., and Murdoch, R. P., The contribution of successful conditioning to abstinence maintenance following covert sensitization (verbal aversion) treatment of alcoholism, IRCS Med. Sci., 5: 167, 1977.

Ellis, A., *Reason and Emotion in Psychotherapy*, Lyle Stuart, New York, N.Y., 1962.

Emmelkamp, P. M. G., *Phobic and Obsessive-Compulsive Disorders: Theory, Research and Practice*, Plenum Press, New York, N.Y., 1982.

Emrick, C. D., A review of psychologically oriented treatment of alcoholism: I. the use and interrelationships of outcome criteria and drinking behavior following treatment, Quart. J. Stud. Alcohol., 35: 523–549, 1974.

Emrick, C. D., A review of psychologically oriented treatment of alcoholism: II. the relative effectiveness of treatment versus no treatment, J. Stud. Alcohol, 36: 88–108, 1975.

Eriksen, L., and Götestam, K. G., Conditioned abstinence in alcoholics: a controlled experiment, Int. J. Addict., 19: 287–294, 1984.

Festinger, L., *A Theory of Cognitive Dissonance*, Stanford University Press, Stanford, Calif., 1959.

Galanter, M., The 'relief effect': a sociobiological model for neurotic distress and large-group therapy, Am J. Psychiat., 135: 588–591, 1978.

Galanter, M., Sociobiology and informal social controls of drinking: findings from two charismatic sects, J. Stud. Alcohol, 42: 64–79, 1981.

Galanter, M., and Buckley, P., Evangelical religion and meditation: psychotherapeutic effects, J. Nerv. Ment. Dis., 166: 685–691, 1978.

Gellman, I. P., *The Sober Alcoholic: An Organizational Analysis of Alcoholics Anonymous*, College and University Press, New Haven, Conn., 1964.

Gerard, D. L., Saenger, G., and Wile, R., The abstinent alcoholic, Arch. Gen. Psychiat., 6: 99–111, 1962.

Glasser, W., *Positive Addiction*, Harper & Row, New York, N.Y., 1976.

Glatt, M. M., The question of moderate drinking despite 'loss of control', Brit. J. Addict., 62: 267–274, 1967.

Goodwin, D. W., The alcoholism of Eugene O'Neill, J.A.M.A., 216: 99–104, 1971.

Goodwin, D. W., *The Writer's Vice*, unpublished manuscript.

Grabowski, J., and O'Brien, C. P., Conditioning factors in opiate use, in N. K. Mello (ed.), *Advances in Substance Abuse*, JAI Press, Inc., Greenwich, Conn., 1981.

Greil, A. L., and Rudy, D. R., Conversion to the world view of Alcoholics Anonymous: a refinement of conversion theory, Qualitative Sociology, 6: 5–27, 1983.

Greil, A. L., and Rudy, D. R., Social cocoons: encapsulation and identity transformation organizations, Sociological Inquiry, 54: 260–278, 1984.

Greil, A. L., and Rudy, D. R., What have we learned from process models of conversion? An examination of ten case studies, Sociological Focus, 17: 305–324, 1984.

Heather, N., and Robertson, I., *Controlled Drinking*, Methuen, London, U.K., 1981.

Heltzer, J. E., Robins, L. N., Taylor, J. R., Carey, K., Miller, R. H., Combs-Orme, T., and Farmer, A., The extent of long-term moderate drinking among alcoholics discharged from medical and psychiatric treatment facilities, New Eng. J. Med., 312: 1678–1682, 1985.

Hodgson, R. J., Rankin, H. J., and Stockwell, T. N., Alcohol dependence and the primary effect, Behav. Research & Ther., 17: 379–387, 1979.

Hoffer, E., *The True Believer*, Harper, New York, N.Y., 1951.

Hoffman, N. G., Harrison, P. A., and Belille, C. A., Alcoholics Anonymous after treatment: attendance and abstinence, Int. J. Addict., 18: 311–318, 1983.

Horowitz, M. J., *States of Mind—Analysis of Change in Psychotherapy*, Plenum, New York, N.Y., 1978.

Hughes, H. E., *The Man From Ida Grove: A Senator's Personal Story*, Chosen Books, Lincoln, Va., 1979.

Huxley, A., *The Doors of Perception*, Harper, New York, N.Y., 1954.

Huxley, A., *Heaven and Hell*, Harper, New York, N.Y., 1958.

Isbell, H., Craving for alcohol, Quart. J. Stud. Alcohol, 16: 38–42, 1955.

Jackson, B., *Afire With Serenity*, Hazelden, Center City, Minn., 1977.

James, W., *Outline of Psychology*, Dover, New York, N.Y., 1950.

James, W. *Principles of Psychology*, Dover, New York, N.Y., 1950.

James, W., *The Varieties of Religious Experience*, Macmillan, New York, N.Y., 1961.

Jellinek, E. M., *The Disease Concept of Alcoholism*, Hillhouse Press, New York, N.Y., 1960.

Jellinek, E. M., Isbell, H., Lundquist, G., Tiebout, H. M., Duchêne, H., Mardones, V., and MacLeod, L. D., The 'craving' for alcohol: a symposium by members of the WHO Expert Committee on Mental Health and Alcohol, Quart. J. Stud. Alcohol, 16: 34–66, 1955.

Johnson, V. E., *I'll Quit Tomorrow*, Harper & Row, New York, N.Y., 1973.

Kaplan, R. F., Cooney, N. L., Baker, L. H., Gillespie, R. A., Meyer, R. E., and Pomerleau, O. F., Reactivity to alcohol-related cues: physiological and subjective responses of alcoholics and nonproblem drinkers, J. Stud. Alcohol, 46: 267–272, 1985.

Kaplan, R. F., Meyer, R. E., and Stroebel, C. F., Alcohol dependence and responsivity to an ethanol stimulus as predictors of alcohol consumption, Brit. J. Addiction, 78: 259–267, 1983.

Keller, M., The oddities of alcoholics, Quart. J. Stud. Alcohol, 33: 147–148, 1972.

Keller, M., On the loss-of-control phenomenon in alcoholism, Brit. J. Addiction, 67: 153–166, 1972.

Kelly, H. H., Attribution theory in social psychology, in D. Levine (ed.), *Nebraska Symposium on Motivation*, Vol. 15, University of Nebraska Press, Omaha, Neb., 1967.

Kelly, H. H., The processes of causal attribution, Amer. Psychologist, 28: 107–128, 1973.

Kirkpatrick, J., *Goodbye Hangovers, Hello Life: Self-help for Women*, Atheneum, New York, N.Y., 1986.

Kissin, B., The role of physical dependence and brain damage in the protracted alcohol abstinence syndrome, Advances in Alcoholism, Vol. 2, No. 11, November 1981.

Knupfer, G., Ex-problem drinkers, in M. Roff, L. N. Robins, and M. Pollock (eds.), *Life History Research in Psychopathology*, University of Minnesota Press, Minneapolis, Minn., 1972.

Kurtz, E., *Not-God: A History of Alcoholics Anonymous*, Hazelden, Center City, Minneapolis, Minn., 1979.

Kurtz, E., Why A.A. works, J. Stud. Alcohol, 43:38–80, 1982.

Kwentus, J., and Major, L. F., Disulfiram in the treatment of alcoholism: a review, J. Stud. Alcohol, 40: 428–446, 1979.

Laverty, S., Aversion therapies in the treatment of alcoholism, Psychosom. Med., 28: 651–666, 1966.

La Barre, W., *The Peyote Cult*, The Shoe String Press, Hamden, Conn., 1964.

Lee, E. H., *Journey Into Nowhere*, Branden Press, Boston, Mass., 1972.

Lemere, F., What happens to alcoholics, Am. J. Psychiatry, 109: 674–676, 1953.

Lifton, R. J., *Thought Reform and the Psychology of Totalism*, Norton, New York, N.Y., 1961.

Litman, G. K., Relapse in alcoholism: traditional and current approaches, in G. Edwards and M. Grant (eds.), *Alcoholism Treatment in Transition*, University Park Press, Baltimore, Md., 1980.

Litman, G. K., Personal meanings and alcoholism survival: translating subjective experience into empirical data, in E. Shepherd and J. P. Watson (eds.), *Personal Meanings*, Wiley, New York, N.Y., 1982.

Litman, G. K., Stapleton, J., Oppenheim, A. N., and Peleg, M., An instrument for measuring coping behaviors in hospitalized alcoholics, Brit. J. Addict., 78: 269–276, 1983.

Litman, G. K., Eiser, J. R., Rawson, N. S. B., and Oppenheim, A. N.,

Differences in relapse precipitants and coping behaviors between alcohol relapsers and survivors, Behav. Res. & Therapy, 17: 89–94, 1979.

Litman, G. K., and Topham, A., Outcome studies on techniques in alcoholism treatment, in M. Galanter (ed.), *Recent Developments in Alcoholism*, Vol. 1, Plenum, New York, N.Y., 1983.

Litman, G. K., Eiser, J. R., Rawson, N. S. B., and Oppenheim, A. N., Toward a typology of relapse: a preliminary report, Drug and Alcohol Dependence, 2: 157–162, 1977.

Litman, G. K., Stapleton, J., Oppenheim, A. N., Peleg, M., and Jackson, P., The relationship between coping behaviours, their effectiveness and alcoholism relapse and survival, Brit. J. Addict., 79: 283–291, 1984.

Litman, G. K., Stapleton, J., Oppenheim, A. N., Peleg, M., and Jackson, P., Situations related to alcoholism relapse, Brit. J. Addict., 78: 381–389, 1983.

Living Sober, Alcoholics Anonymous World Services, Inc., New York, N.Y., 1975.

Lofland, J., "Becoming a world-saver" revisited, Am. Behavioral Scientist, 20: 805–818, 1977.

Lofland, J., and Stark, R., Becoming a world-saver; a theory of conversion to a deviant perspective, Am. Sociological Review, 30: 862–874, 1965.

London, J., *John Barleycorn*, Grosset & Dunlap, New York, N.Y., 1913.

Low, A. A., *Mental Health Through Will Training*, Christopher Publishing House, West Hanover, Mass., 1950.

Ludwig, A. M., Altered states of consciousness, Arch. Gen. Psychiat., 15: 225–234, 1966.

Ludwig, A. M., The formal characteristics of therapeutic insight, Am. J. Psychotherapy, 20: 305–318, 1966.

Ludwig, A. M., On and off the wagon: reasons for drinking and abstaining by alcoholics, Quart. J. Stud. Alcohol, 33: 91–96, 1972.

Ludwig, A. M., The irresistible urge and the unquenchable thirst for alcohol, in M. E. Chafetz (ed.), *Proceedings of the Fourth Annual Alcoholism Conference of the National Institute on Alcohol Abuse and Alcoholism: Research, Treatment and Prevention*, Government Printing Office, Washington, D.C., 1975.

Ludwig, A. M., Why alcoholics drink, in B. Kissin and H. Begleiter (eds.), *The Biology of Alcoholism. Vol. 6: The Pathogenesis of Alcoholism: Psychosocial Factors*, Plenum, New York, N.Y., 1983.

Ludwig, A. M., Cognitive processes associated with 'spontaneous' recovery from alcoholism, J. Stud. Alcohol, 46: 53–58, 1985.

Ludwig, A. M., Pavlov's 'bells' and alcohol craving, Addictive Behaviors, 11: 87–91, 1986.

Ludwig, A. M., Cain, R. B., Wikler, A., and Bendfeldt, F., Physiologic and situational determinants of drinking behavior, in M. M. Gross (ed.), *Alcohol Intoxication and Withdrawal: Experimental Studies*, Vol. 3, Plenum, New York, N.Y., 1977.

Ludwig, A. M., and Levine, J., A controlled comparison of five brief treatment techniques employing LSD, hypnosis, and psychotherapy, Am. J. Psychotherapy, 19: 417–435, 1965.

Ludwig, A. M., Levine, J., and Stark, L. H., *LSD and Alcoholism: A Clinical Study of Treatment Efficacy*, Charles C. Thomas, Springfield, Ill., 1970.

Ludwig, A. M., and Lyle, W. H., Jr., The experimental production of narcotic drug effects and withdrawal symptoms through hypnosis, Int. J. Clin. Exper. Hypnosis, 12: 1–17, 1964.

Ludwig, A. M., and Othmer, E., The medical basis of psychiatry, Am. J. Psychiatry, 134: 1087–1092, 1977.

Ludwig, A. M., and Stark, L. H., Alcohol craving: subjective and situational aspects, Quart. J. Stud. Alcohol, 35: 899–905, 1974.

Ludwig, A. M., and Stark, L. H., "Arousal" and alcoholism: Psychophysiological responses to alcohol, in M. M. Gross (ed.), *Alcohol Intoxication and Withdrawal*, Plenum, New York, N.Y., 1975.

Ludwig, A. M., and Wikler, A., Craving and relapse to drink, Quart. J. Stud. Alcohol, 35: 108–130, 1974.

Ludwig, A. M., Wikler, A., and Stark, L. H., The first drink: psychophysiological aspects of craving, Arch. Gen. Psychiatry, 30: 539–547, 1974.

Lynch, J. J., Fertziger, A. P., Teitelbaum, H. A., Cullen, J. W., and Gantt, W. H., Pavlovian conditioning of drug reactions: some implications for problems of drug addiction, Conditional Reflex, 8: 211–223, 1973.

MacAndrew, C., and Edgerton, R. B., *Drunken Comportment: A Social Explanation*, Aldine, Chicago, Ill., 1969.

Madill, M. F., Campbell, D., Laverty, S. G., Sanderson, R. E., and Vandewater, S. L., Aversion treatment of alcoholics by succinylcholine-induced apneic paralysis, Quart. J. Stud. Alcohol, 27: 483–509, 1966.

Madsen, W., *The American Alcoholic*, Charles C. Thomas, Springfield, Ill., 1974.

Mahoney, M. J., *Cognitive and Behavior Modification*, Ballinger, Cambridge, Mass., 1974.

Marlatt, G. A., Craving for alcohol, loss of control, and relapse: a cognitive-behavioral analysis, in P. E. Nathan, G. A. Marlatt, and T. Loberg (eds.), *Alcoholism: New Directions in Behavioral Research and Treatment*, Plenum, New York, N.Y., 1978.

Marlatt, G. A., A cognitive-behavioral model of the relapse process, in N. A.

Krasnegor (ed.), *Behavioral Analysis and Treatment of Substance Abuse*, NIDA Research Monograph 25, June 1979.

Marlatt, G. A., Cognitive assessment and intervention procedures for relapse prevention, in G. A. Marlatt and J. R. Gordon (eds.), *Relapse Prevention*, Guilford Press, New York, N.Y., 1985.

Marlatt, G. A., Coping and substance abuse: implications for research, prevention, and treatment, in S. Shiffman and T. A. Wills (eds.), *Coping and Substance Use*, Academic Press, New York, N.Y., 1985.

Marlatt, G. A., and George, W. H., Relapse prevention: introduction and overview of the model, Brit J. Addict., 79: 261–273, 1984.

Marlatt, G. A., and Gordon, J. R., (eds.), *Relapse Prevention*, Guilford Press, New York, N.Y., 1985.

Marlatt, G. A., and Rohsenow, D. J., The think-drink effect, Psychology Today, 15: 60–69, 93, 1981.

Mathew, R. J., Claghorn, J. L., and Largen, J., Craving for alcohol in sober alcoholics, Am. J. Psychiatry, 136: 603–606, 1979.

Maultsby, M. C., Jr., *A Million Dollars for Your Hangover*, Rational Self-Help Books, Lexington, Ky., 1978.

Maultsby, M. C., Jr., *Rational Behavior Therapy*, Prentice-Hall, Englewood Cliffs, N.J., 1984.

Maxwell, M. A., *The A.A. Experience*, McGraw-Hill, New York, N.Y., 1984.

McCance, C., and McCance, P. F., Alcoholism in North-East Scotland: its treatment and outcome, Brit. J. Psychiatry, 115: 189–198, 1969.

Meichenbaum, D., *Cognitive-Behavior Modification: An Integrative Approach*, Plenum, New York, N.Y., 1977.

Mello, N. K., Some aspects of the behavioral pharmacology of alcohol, in D. H. Efron, J. O. Cole, J. Levine, and J. R. Wittenborn, (eds.), *Psychopharmacology: a Review of Progress 1957–1967*, U.S. Government Printing Office, Washington, D.C., 1968.

Mello, N. K., Behavioral studies in alcoholism, in B. Kissin and H. Begleiter (eds.), *Biology of Alcoholism*, Vol. 2, Plenum, New York, N.Y., 1972.

Mello, N. K., McNamee, H. B., and Mendelson, J. H., Drinking patterns of chronic alcoholics: gambling and motivation for alcohol, Psychiat. Research Reports, 24: 83–118, 1968.

Mello, N. K., and Mendelson, J. H., Operant analysis of drinking patterns of chronic alcoholics, Nature, 206: 43–46, 1965.

Mello, N. K., and Mendelson, J. H., Drinking patterns during work-contingent and non-contingent alcohol acquisition, in N. K. Mello and J. H. Mendelson (eds.), *Recent Advances in Studies of Alcoholism*, National Institute on Alcohol Abuse and Alcoholism, Rockville, Md., 1971.

Merbaum, M., and Rosenbaum, M., Self-control theory and technique in the modification of smoking, obesity, and alcohol abuse, Clin. Behav. Ther. Review, 2: 1–20, 1980.

Merry, J., The 'loss of control' myth, Lancet, 1: 1257–1258, 1966.

Meyer, R. E., and Mirin, S. M., *The Heroin Stimulus: Implications for a Theory of Addiction*, Plenum, New York, N.Y., 1979.

Miller, P.M., Theoretical and practical issues in substance abuse assessment and treatment, in W. R. Miller (ed.), *The Addictive Behaviors*, Pergamon Press, Oxford, U.K., 1980.

Musgraves, D., and Balsiger, D., *One More Time*, Bethany House, Minneapolis, Minn., 1974.

Neubeurger, O. W., Matarazzo, J. D., and Schmitz, R. E., One year follow-up of total abstinence in chronic alcoholic patients following counterconditioning, Alcoholism: Clin. & Exper. Research, 4: 306–312, 1980.

O'Brien, C. P., Experimental analysis of conditioning factors in human narcotic addiction, Pharmacol. Reviews, 27: 533–543, 1975.

O'Brien, C. P., O'Brien, T. J., Mintz, J., and Brady, J. P., Conditioning of narcotic abstinence symptoms in human subjects, Drug and Alcohol Dep., 1: 115–123, 1975.

Only, M., *High: A Farewell to the Pain of Alcoholism*, Prentice-Hall, Englewood Cliffs, N.J., 1974.

Overton, D. A., State dependent learning produced by alcohol and its relevance to alcoholism, in B. Kissin and H. Begleiter (eds.), *Biology of Alcoholism. Vol. II: Physiology and Behavior*, Plenum, New York, N.Y., 1972.

Pahnke, W. N., The contribution of the psychology of religion to the therapeutic use of psychedelic substances, in H. A. Abramson (ed.), *The Use of LSD in Psychotherapy and Alcoholism*, Bobbs-Merrill, Indianapolis, 1966.

Pahnke, W. N., and Richards, W. A., Implications of LSD and experimental mysticism, in C. T. Tart (ed.), *Altered States of Consciousness*, Wiley, New York, N.Y., 1969.

Parratt, L. W., *My Alcoholic Virgin*, Vantage Press, New York, N.Y., 1972.

Peele, S., *How Much Is Too Much: Healthy Habits or Destructive Addictions*, Prentice-Hall, Englewood Cliffs, N.J., 1981.

Peele, S., Out of the habit trap, Am. Health, 2: 42–47, 1983.

Peele, S., *The Meaning of Addiction: Compulsive Experience and Its Interpretation*, Lexington Books, Lexington, Mass., 1985.

Peele, S., and Brodsky, A., *Love and Addiction*, Taplinger Publishing Co., New York, N.Y., 1975.

Perri, M. G., Self-change strategies for control of smoking, obesity, and problem drinking, in S. Shiffman and T. A. Wills (eds.), *Coping and Substance Use*, Academic Press, New York, N.Y., 1985.

Polich, J. M., Armor, D. J., and Braiker, H. B., *The Course of Alcoholism: Four Years After Treatment*, Wiley, New York, N.Y., 1981.

Proudfoot, W., and Shaver, P., Attribution theory and the psychology of religion, J. Scientific Study of Religion, 14: 317–330, 1976.

Rachman, S. J., and Hodgson, R. F., *Obsessions and Compulsions*, Prentice-Hall, Englewood Cliffs, N.J., 1980.

Rankin, H. J., Hodgson, R. J., and Stockwell, T., The behavioral measurement of alcohol dependence, Brit. J. Addiction, 75: 43–47, 1980.

Redgrove, P., *The Beekeepers*, Routledge and Kegan Paul, London, U.K., 1980.

Richardson, J. T., and Stewart, M., Conversion process models and the Jesus movement, Am. Behav. Scient., 20: 819–838, 1977.

Riddell, P., *I Was an Alcoholic: The Story of a Curse*, Victor Gollancz Ltd., London, U.K., 1955.

Rudy, D. R., Slipping and sobriety: the functions of drinking in Alcoholics Anonymous, J. Stud. Alcohol, 41: 727–732, 1980.

Rudy, D. R., *Becoming Alcoholic*, Southern Illinois University Press, Carbondale, Ill., 1986.

Salzman, L., The psychology of religious and ideological conversion, Psychiatry, 16: 177–187, 1953.

Sargant, W., *Battle for the Mind*, Doubleday, Garden City, N.Y., 1957.

Saunders, W. M., and Kershaw, P. W., Spontaneous remission from alcoholism: a community study, Brit. J. Addiction, 74: 251–265, 1979.

Schachter, S., Recidivism and self-cure of smoking and obesity, Am. Psychologist, 37: 436–444, 1982.

Schachter, S., and Singer, J. E., Cognitive, social and physiological determinants of emotional state, Psychological Rev., 69: 379–399, 1962.

Shiffman, S., Coping with temptations to smoke, in S. Shiffman and T. A. Wills (eds.), *Coping and Substance Use*, Academic Press, New York, N.Y., 1985.

Shiffman, S., and Wills, T. A. (eds.), *Coping and Substance Use*, Academic Press, New York, N.Y., 1985.

Sideroff, S. I., and Jarvik, M. E., Conditioned responses to a videotape showing heroin-related stimuli, Int. J. Addict., 15: 529–536, 1984.

Siegler, M., and Osmond, H., *Models of Madness, Models of Medicine*, Macmillan, New York, N.Y., 1974.

Simmonds, R. B., Conversion or addiction: consequences of joining a Jesus movement group, Am. Behav. Sci., 20: 909–924, 1977.

Smart, R. G., Spontaneous recovery in alcoholics: a review and analysis of the available research, Drug Alcohol Depend., 1: 277–285, 1975/76.

Solomon, R. L., An opponent-process theory of acquired motivation: IV. the affective dynamics of addiction, in J. Maser and M. E. P. Seligman (eds.), *Psychopathology: Experimental Models*, W. H. Freeman, San Francisco, Calif., 1977.

Solomon, R. L., The opponent-process theory of acquired motivation: the costs of pleasure and the benefits of pain, Am. Psychologist, 35: 691–712, 1980.

Solomon, R. L., and Corbit, J. D., An opponent-process theory of motivation: temporal dynamics of affect, Psychol. Rev., 81: 119–145, 1974.

Stall, R. D., An examination of spontaneous remission from problem drinking in the Bluegrass area of Kentucky, Master's Thesis, University of Kentucky, 1979.

Starbuck, E. D., *The Psychology of Religion*, Charles Scribner's Sons, New York, N.Y., 1900.

Stockwell, T. R., Hodgson, R. J., Rankin, R.J., and Taylor, C., Alcohol dependence, beliefs and the primary effect, Behav. Res. & Therap., 20: 513–522, 1982.

Straus, R., Types of alcohol dependence, in B. Kissin and H. Begleiter (eds.), *The Biology of Alcoholism. Vol. 6: The Pathogenesis of Alcoholism: Psychosocial Factors*, Plenum, New York, N.Y., 1983.

Straus, R., Alcohol and alcohol problems research: the United States, Brit. J. Addiction, 81: 313–323, 1986.

Ternes, J. W., O'Brien, C. P., et al., Conditioned drug responses to naturalistic stimuli, National Institute of Drug Abuse Research Monograph Series, 27: 282–288, 1979.

Terry, R., *The Long Suffering*, Paulist Press, New York, N.Y., 1976.

Thomsen, R., *Bill W.*, Harper & Row, New York, N.Y., 1975.

Tiebout, H. M., The act of surrender in the therapeutic process, distributed by the National Council on Alcoholism, New York, N.Y.

Tiebout, H. M., Conversion as a psychological phenomenon, distributed by the National Council on Alcoholism, New York, N.Y.

Tiebout, H. M., Therapeutic mechanisms of Alcoholics Anonymous, in *Alcoholics Anonymous Comes of Age: A Brief History of A.A.*, Alcoholics Anonymous World Services, Inc., 1957.

Tiebout, H. M., Surrender versus compliance with therapy: with special reference to alcoholism, Quart. J. Stud. Alcohol, 14: 58–68, 1959.

Tokar, J. T., Brunse, A. J., Stefflre, V. J., Napior, D. A., and Sodergren, J. A., Emotional states and behavioral patterns in alcoholics and nonalcoholics, Quart. J. Stud. Alcohol, 34: 133–143, 1973.

Tournier, R. E., Alcoholics Anonymous as treatment and as ideology, J. Stud. Alcohol, 40: 230–239, 1979.

Trice, H. M., and Roman, P. M., Sociopsychological predictors of affiliation with Alcoholics Anonymous: a longitudinal study of "treatment success," Social Psychiatry, 5: 51–59, 1970.

Tuchfeld, B. S., Spontaneous remission in alcoholics: empirical observations of theoretical implications, J. Stud. Alcohol, 42: 626–641, 1981.

Tuchfeld, B. S., and Marcus, S. H., The resolution of alcohol-related problems: in search of a model, J. Drug Issues, 14: 151–159, 1984.

Twelve Steps and Twelve Traditions, Alcoholics Anonymous Publishing Co., New York, N.Y., 1983.

Vaillant, G. E., *The Natural History of Alcoholism*, Harvard University Press, Cambridge, Mass., 1983.

Vaillant, G. E., and Milofsky, E. S., Natural history of male alcoholism: paths to recovery, Arch. Gen. Psychiatry, 39: 127–133, 1982.

Victor, M., and Abrams, R. D., The effect of alcohol on the nervous system, in H. H. Merritt and C. C. Hare (eds.), *Metabolic and Toxic Diseases of the Nervous System*, Williams & Wilkins, Baltimore, Md., 1953.

Wechsler, H., The self-help organization in the mental health field: Recovery, Inc., a case study, J. Nerv. Ment. Dis., 130: 297–314, 1960.

Wheelis, A., *How People Change*, Harper & Row, New York, N.Y., 1973.

Wholey, D., *The Courage to Change*, Houghton-Mifflin, Boston, Mass., 1984.

Wiens, A. N., and Menustik, C. E., Treatment outcome and patient characteristics in an aversion therapy program for alcoholism, Am. Psychologist, 38: 1089–1096, 1983.

Wikler, A., Recent progress in research on the neurophysiological basis of morphine addiction, Amer. J. Psychiat., 105: 329–338, 1948.

Wikler, A., A psychodynamic study of a patient during experimental self-regulated re-addiction to morphine, Psychiat. Quart., 26: 270–293, 1952.

Wikler, A., A rationale of the diagnosis and treatment of addictions, Conn. St. Med. J., 19: 560–569, 1955.

Wikler, A., On the nature of addiction and habituation, Brit. J. Addiction, 57: 73–79, 1961.

Wikler, A., Conditioning factors in opiate addiction and relapse, in D. I. Wilner and G. G. Kassenbaum (eds.), *Narcotics*, McGraw-Hill, New York, N.Y., 1965.

Wikler, A., Interaction of physical dependence and classical and operant conditioning in the genesis of relapse, Research Publication of the Association of Nervous and Mental Disease, 46: 280–287, 1968.

Wikler, A., Some implications of conditioning theory for problems of drug abuse, Behav. Science, 16: 92–97, 1971.

Wikler, A., Dynamics of drug dependence: implications of a conditioning theory for research and treatment, Arch. Gen. Psychiat., 28: 611–616, 1973.

Wikler, A., Sources of reinforcement for drug using behavior: a theoretical formulation, Proceedings of the 5th International Congress of Pharmacology, 1: 18–30, 1973.

Wikler, A., Pescor, F. T., Miller, D., and Norrell, H., Persistent potency of a secondary (conditioned) reinforcer following withdrawal of morphine from physically dependent rats, Psychopharmacologia, 20: 103–117, 1971.

Wikler, A., and Pescor, F. T., Classical conditioning of morphine-addicted rats, Psychopharmacologia, 10: 255–284, 1967.

Williams, T., Cat on a Hot Tin Roof, in *The Theatre of Tennessee Williams, Vol. III*, New Directions Books, New York, N.Y., 1971.

Wilson, B., Basic concepts of Alcoholics Anonymous, Address presented to the Medical Society of the State of New York, Section on Neurology and Psychiatry, Annual Meeting, New York, May 1944.

Wilson, B., The society of Alcoholics Anonymous, presented at the 105th Annual Meeting of the American Psychiatric Association, Montreal, Quebec, May 1949.

Wilson, B., Alcoholics Anonymous—beginnings and growth, presented to the New York City Medical Society on Alcoholism, April 28, 1958.

Wilson, W. P., Mental health benefits of religious salvation, Dis. Nerv. Syst., 33: 382–386, 1972.

Zaehner, R. C., *Mysticism: Sacred and Profane*, Oxford University Press, New York, N.Y., 1961.

Index